CHEMICAL INFORMATION

CHEMICAL INFORMATION

A Practical Guide to Utilization

Second Revised and Enlarged Edition

Yecheskel Wolman

Department of Organic Chemistry
The Hebrew University of Jerusalem

A Wiley-Interscience Publication

JOHN WILEY & SONS
Chichester · New York · Brisbane · Toronto · Singapore

Library of Congress Cataloging-in-Publication Data:

Wolman, Yecheskel, 1935–
  Chemical information.

  Includes index.
  1. Chemical literature.   2. Chemistry—Information
services.   1. Title.
QD8.5.W64 1988   540'.72   87-23119
ISBN 0 471 91704 4

British Library Cataloguing in Publication Data:

Wolman, Yecheskel
  Chemical Information: a practical guide
  to utilization.—2nd Rev. and enl. ed.
  1. Chemistry—Information services
  2. Chemistry—Bibliography
  1. Title
  540'.7   QD8.3

  ISBN 0 471 91704 4

Printed in Great Britain by St Edmundsbury Press,
Bury St Edmonds.

To Bev

Contents

Preface to the Second Edition

This second edition of *Chemical Information, A Practical Guide to Utilization*, is the best demonstration of the rapid changes and developments which took place in the area of chemical information during the last five or six years. Sources that were used not long ago have been replaced by others, tools that were a must only yesterday are already forgotten today, more and more paper products are being replaced (or supplemented) by computer products. Many of today's tools will probably be changed by others in the near future.

The changes which occurred during the earlier and mid 1980s are reflected in this book. Online searching, which became a routine tool during that period, is not treated separately but is discussed through the book parallel to the various manual searching tools and sources. The basic principles and applications of expert systems, systems which are going to become more and more important in chemical information in the near future, are discussed in a separate new chapter. Another chapter is devoted to trends and perspectives in chemical information, trying to predict some of the tools and methods which will be used in the not too distant future.

Another development which took place during the early eighties is the introduction of up-to-date formal and/or informal courses in chemical information to the chemistry curricula of the undergraduate and/or graduate chemistry major students. As a matter of fact, this second edition of the book reflects the changes which took place at our teaching programme of chemical information at the Institute of Chemistry of the Hebrew University of Jerusalem.

I am indebted to my friends, colleagues and students at the Hebrew University who contributed helpful comments and remarks. Special thanks go to Ms. A. Chapman and Dr. B. A. Gore, Derwent Publication Co.; Dr. D. F. Chudos, Distributed Chemical Graphics Inc.; Ms. A. M. Cunningham, Ms. B. Lawlor and Dr. M. Main, Institute of Scientific Information; Mr. R. C. Dana, Mr. E. P. Donnell, Dr. G. P. Platau, Dr. R. Turkel and Dr. B. Zahn, Chemical Abstract Service; Dr. S. Fivozinsky, Office of Standard Reference Data, National Bureau of Standards, US Department of Commerce; Prof. E. Fluck, Director of the Gmelin Institute; Dr. S. R. Heller, USDA, ARS; Prof. R. Luckenbach, Director of the Beilstein Institute; and Mr. R. Stileman, Chapman and Hall. They provided me with an insight into their specialities and helped me to gain a general overview of various aspects and topics of chemical information.

Yecheskel Wolman

CHAPTER 1

Introduction

Scientists in all disciplines have the need today, more than ever, for better communication with fellow scientists. Communication is needed in order to enable one to keep up to date with the latest developments in one's field of research, to obtain specific data (e.g. chemical, biological, physical), to check the validity of a new theory, etc.

Effective and efficient communication among chemists today is possible only via an intermediate medium—the primary chemical literature (such is the case in any other area of science). Even this kind of communication, with well over 400,000 articles and about 100,000 patents published annually (not to mention other written forms: reports, deposited documents, conference abstracts and/or proceedings, theses, books), is hard to follow. It is interesting that although there are about 14,000 scientific journals that publish original chemical research results, 25% of all published articles appear in only 90 journals (e.g. in 1984 about 1800 scientific articles were published in the *Journal of the American Chemical Society*—about 0.45% of the 400,000 papers published annually), 50% in 450 journals, and 75% in 1600 journals.

In order to keep up to date in one's own field of research, and to become acquainted with new developments and trends in science, one has to cover the current literature. How can this be done? The individual chemist is not able to read all or even most of the chemical journals. Most journals today are in some specialized field, which limits the number of journals one is interested in, but nevertheless the average chemist reads a very small number of journals (usually 10–25). Some of them are of the more general type while others are in his own specific area of research. For access to news and developments not reported in these journals, the chemist refers to the secondary literature (lists of titles, abstracts, information sheets, etc.), which leads him to the original literature on topics of particular interest. But how is this done in practice?

The situation is much more complicated when one is looking for a specific data (e.g. the temperature at which the vapour pressure of ethanol is 4 atm), trying to formulate an answer based on fundamental information available (e.g. the structure–function relationship of vasopressin), or just looking for general background material in a new research area. How is the required information to be found?

All research results have been reported in the primary scientific literature (primary journals, reports, deposit documents, theses, proceedings and/or abstracts of scientific meetings, patent applications) for over 200 years. Data on the physical, biological or chemical properties of a substance, method of preparation of a compound, etc., may be used by the chemist today even if it was first reported many years ago. All that is required is that he be able to locate the required information, however long ago it was published.

An example of important available information that was difficult to locate is the data concerning the German synthetic oil programme. Although it was well known that during the Second World War the German war machine was run almost exclusively on coal-derived gasoline, there was no reference to the German technology in the scientific literature for over three decades. This was not because of lack of documentary information; plant diagrams (at the end of the War 25 oil-from-coal plants existed), patents, data on the catalysts and additives used, as well as environmental controls data were available. The problem was that the information was not reported in any of the common sources and was difficult to locate.

Correct usage of the chemical literature and other available information sources requires considerable effort. However, the benefits gained by correct usage make these extra efforts worth while.

There are a large number of books devoted to the chemical literature and its usage, and a few should be mentioned:

H.M. Woodburn, *Using the Chemical Literature. A Practical Guide*, Marcel Dekker, New York, 1974, 302pp.

R.T. Bottle (Editor), *Use of Chemical Literature*, 3rd ed., Butterworths, London, 1979, 320pp.

A. Antony, *Guide to Basic Information Sources in Chemistry*, Jeffrey Norton Publishers, New York, 1979, 219pp.

H. Skolnik, *The Literature Matrix of Chemistry*, Wiley-Interscience, New York, 1982, 297pp.

M.G. Mellon, *Chemical Publications, Their Nature and Use*, 5th ed. McGraw-Hill, New York, 1982, 395pp.

J. Ash, P. Chubb, S. Ward, S. Welford and P. Willett, *Communication, Storage and Retrieval of Chemical Information*, Ellis Horwood, Chichester, 1985, 297pp.

R.E. Maizel, *How to Find Chemical Information. A Guide for Practicing Chemists, Educators and Students*, 2nd ed., Wiley-Interscience, New York, 1986, 412pp.

These books include discussion of primary and secondary sources. Some concentrate more heavily on the various references tools, such as abstracts, indexes, handbooks and dictionaries, whereas others put the emphasis on the various types of primary literature—journals, reports, theses, and patents.

Some are more comprehensive than others but none emphasize real search problems and their solutions.

This book is intended to teach and demonstrate the practical use of the chemical literature. It will show and teach the chemist (whether he is a graduate or an undergraduate student, or whether he is dealing with academic or industrial research, scientific administration, sales, teaching or consulting) how to keep up to date in his own area of interest and will show him the best way to locate any information he requires using various paper, microform and electronic tools and products. This will be done by giving real search examples from various areas of chemistry. Some of the information could be located by using well known sources, whereas the remainder could be found by other means. The different ways of obtaining the required data will be compared and the various means and methods used for the purpose will be critically evaluated.

CHAPTER 2

The Scientific Journal

The scientific article is and will be (at least for the foreseeable future) the most important and essential document to scientists, as it is the basic unit of the scientific journal. The scientific journal itself is public, formal, and the most organized channel of communication among the scientific community. It is public as anybody can submit a manuscript to a scientific journal, and the journal itself is available to anyone either by personal subscription or through the library. It is formal as articles are examined, reviewed and (if there is a need) revised to near perfection. The organization is based on the acceptance of articles on the basis of their scientific merits; thus those articles report flawlessly research results which are a part of the scientific progress.

The formal as well as the organizational aspects of the scientific journal are taken care of by fellow scientists (peer review). This is understandable, as only people that have proven themselves in their scientific discipline could judge and decide about the novelty, quality and importance of a contribution to this specific field.

2.1. The scientific article

There are four types of journal articles—a full paper, a note (or a short paper), a communication (known also as a preliminary communication or a letter to the editor), and a review.

A full paper presents new important data or provides a new and fresh approach to an established subject. It usually consists of an informative, concise abstract, which summarizes the principle findings of the work, an introduction, which clearly states the problems, the background of the work and the approach of the author, a detail experimental section, which gives information about materials and non-standard apparatus used and a full description of how the work was done, a summary of the results, and a discussion of their significance. A note or a short paper is of course shorter than a full paper and describes more limited findings. Notes usually do not have an abstract—the emphasis is on the experimental section, while the introduction, results and discussion are short. Communications are preliminary reports of results of special significance and urgency that are given expedited publication. They consist mainly of results and discussion sections and have only a very short introduction and minimum experimental data; some communications are followed later by a full paper. The authors often have to justify the submission of a scientific contribution for publication as a preliminary communication in a covering letter to the journal editor. Review articles usually do not contain new experimental results nor experimental details (reviews published in *Synthesis* or *Organic Reactions* are among the exceptions and do contain a few experimental examples). Reviews do correlate and integrate results, relevant to the subject in question, from various publications. One of the most important functions of a review is to serve as a guide to the original literature. When conducting a retrospective search, it is advisable first to examine reviews dealing with the subject before reading the original papers. Reviews can be divided into three categories. First, a comprehensive coverage of the subject up to the time of writing (e.g. *Chemical Reviews, Organic Reactions*); second, coverage of a subject for a limited period of time, usually a year or two [e.g. *The Specialist Periodical Reports Series* published by The Royal Society of Chemistry (RSC) (formerly the Chemical Society)]; and third, reviews written by people working intensively in the field and whose contribution to it is well recognized, and in which the emphasis is on the author's own views and contributions (e.g. *Accounts of Chemical Research*). Reviews are published not only in journals but also in a large number of serials prefixed by: *Progress in...*, *Advances in...*, *Fortschritte...*, etc., as well as by other serial publications (e.g. *Organic Reactions, Survey of Progress in Chemistry, Topics in Current Chemistry*).

Very few journals publish all four types of articles. Few publish three types (e.g. *Journal of Medicinal Chemistry* and *Journal of Organic Chemistry* publish full papers, notes, and communications). Some publish only papers and do not differentiate between a full paper and a note (e.g. *Journal of the Chemical Society Perkin Transactions I* and *II*), whereas others publish only full papers and communications (e.g. *Journal of the American Chemical Society*). There are journals that publish only communications (e.g. *Journal of the Chemical Society Chemical Communications, Tetrahedron Letter*). Some journals publish review articles and notes (e.g. *Synthesis, Angewadte Chemie*) while others publish reviews and papers without differentiation between a full paper and a note (e.g. *Tetrahedron*).

2.2. Publication of scientific journals

Who publishes the scientific journals and what are the publishers' interest in their publication?

One of the main functions of a scientific society is the provision of good scientific journals. Indeed, most of the prestige journals are published by scientific societies. The American Chemical Society (ACS) publishes 22 high-quality journals (Table 2.1); furthermore, the Chemical Abstract Service

Table 2.1 Journals published by the ACS

Accounts of Chemical Research
Analytical Chemistry
Biochemistry
Chemical Reviews
Chemical and Engineering News
Chemtech
Energy and Fuels
Environmental Science and Technology
Industrial and Engineering Chemistry
InorganicChemistry
Journal of the American Chemical Society
Journal of Chemical and Engineering Data
Journal of Chemical Education
Journal of Chemical Information and Computer Science
Journal of Medicinal Chemistry
Journal of Organic Chemistry
Journal of Physical Chemistry
Journal of Physical and Chemical Reference Data
Langmuir
Macromolecules
Organometallics

(CAS) which is the publisher of *Chemical Abstracts* (*CA*) is a division of the American Chemical Society.

Among the various activities of governmental agencies dealing with research and development, the publication of primary scientific journals is included, e.g. the National Research Council of Canada (NRCC) is the publisher of 12 scientific journals (Table 2.2), 6 of which are of interest to chemists.

Table 2.2 Journals published by the NRCC

Canadian Geotechnical Journal
Canadian Journal of Biochemistry and Cell Biology
Canadian Journal of Botany
Canadian Journal of Chemistry
Canadian Journal of Civil Engineering
Canadian Journal of Earth Science
Canadian Journal of Forest Research
Canadian Journal of Genetics and Cytology
Canadian Journal of Microbiology
Canadian Journal of Physics
Canadian Journal of Physiology and Pharmacology
Canadian Journal of Zoology

Publishing scientific journals is still a profitable business today, and indeed various commercial publishing houses publish a large number of journals in various fields of science (e.g. natural sciences, medicine, social sciences).

Many chemistry and chemistry-related journals are published by John Wiley and Sons, some of which are listed in Table 2.3.

Table 2.3 Some chemistry and chemistry-related
journals published by John Wiley and Sons

Fire and Materials
Flavour and Fragrance Journal
International Journal of Chemical Kinetics
International Journal of Quantum Chemistry
International Petroleum Abstracts
Journal of Applied Polymer Science
Journal of Applied Toxicology
Journal of Chemometrics
Journal of Computational Chemistry
Journal of Molecular Electronics
Journal of Polymer Science
Journal of Raman Spectroscopy
Journal of the College of Science
Journal of Thermal Analysis
Magnetic Resonance in Chemistry
Mass Spectroscopy Reviews
Organic Mass Spectrometry
Surface and Interface Analysis
Thermal Analysis Abstracts
X-Ray Spectroscopy

In 1665 the first scientific journal, *Philosophical Transactions of the Royal Society*, was published (6 May 1665). The first chemical journal, *Chemisches Journal für die Freunde der Naturlehre*, appeared just over 100 years later (1778). At the begining of the 19th Century, about 100 scientific journals (many of them dealing, in whole or in part, with chemistry) were in circulation. Since then, the number of scientific journals has constantly increased. Over the years some of the journals stopped publication (e.g. *Popular Science News and Boston Journal of Chemistry, Zeitschrift für Chemie*) while others changed names and/or split into sections (e.g. *Berichte der Deutschen Chemischen Gesellschaft* changed its name after 77 years of publication to *Chemische Berichte*, the publication history of the *Journal of the Chemical Society* is much more complicated—see Table 2.4). Some journals started with one or two volumes a year and now publish dozens of volumes a year (e.g. *Journal of Organometallic Chemistry*, started with two volumes in 1964, and published 19 volumes in 1984; *Biophysica Biochimica Acta* started with one volume in 1947 and published 12 volumes in 1963 and 42 volumes in 1985).

The increase in the number of volumes as well as the subdivision of the scientific journals is caused by the ever-increasing volume of scientific knowledge. A primary journal is intended to cover a well defined subject area. During the years, as this area grows and develops, so also grow the number

Table 2.4   The publishing history of the journal of the Chemical Society

1841–1843	*Proceedings of the Chemical Society of London*
1841–1843	*Memoirs of the Chemical Society of London*
1843–1848	*Memoirs and Proceedings of the Chemical Society of London*
1849–1862	*Quarterly Journal of the Chemical Society of London*
1862–1965	*Journal of the Chemical Society*
1957–1963	*Proceedings of the Chemical Society*
1964–1968	*Chemical Communications*
1966–1971	*Journal of the Chemical Society A*
	Journal of the Chemical Society B
	Journal of the Chemical Society C
1969–1971	*Journal of the Chemical Society D*
1972	*Journal of the Chemical Society, Dalton Transactions*
	Journal of the Chemical Society, Perkin Transactions I
	Journal of the Chemical Society, Perkin Transactions II
	Journal of the Chemical Society, Chemical Communications
	Journal of the Chemical Society, Faraday Transactions I
	Journal of the Chemical Society, Faraday Transactions II

of specialities within the area, the number of scientists working in the field, and the number of scientific papers within each discipline per year. All this leads to an increase in the size of the journal volume followed by a decrease in the overall relevance of the journal to the individual reader. In order not to lose authors and readers the journal eventually must split into sections.

One has to keep in mind that there are journals that publish one un-numbered volume per year (e.g. *Chemistry and Industry, Chemistry Letters*); others started with numbered volumes and now publish one unnumbered volume per year (e.g. *Journal of the Chemical Society*), or started with un-numbered volumes and changed into numbered ones (e.g. *Tetrahedron Letters*). Most of the journals appear once or twice a month but there are also many weekly (e.g. *Science, Nature*), bimonthly (e.g. *Journal of Research of the National Bureau of Standards*), and trimonthly (e.g. *Quarterly Review of Biophysics*) publications.

There are a limited number of general chemistry journals, dealing with the various aspects of chemistry (e.g. *Journal of the American Chemical Society, Helvetica Chimica Acta, Bulletin of the Chemical Society of Japan, Nou-veau Journal de Chimie*). There are journals that are devoted to a branch of chemistry (e.g. *Tetrahedron, Tetrahedron Letters, Journal of Organic Chemistry, Angewandte Chemie, Synthesis, Journal of the Chemical Society Perkin Transactions* are just a few of the journals devoted to organic chemistry). Some journals deal with a defined narrow area of chemistry (e.g. *Hetero-cycles, Journal of Heterocyclic Chemistry, Chemistry of Heterocyclic Com-pounds* are three journals devoted to heterocyclic chemistry). There are also journals devoted to interdisciplinary subjects (e.g. the study of origins of life is covered by a number of journals such as *Origins of Life, Biosystems, Journal of Molecular Evolution*).

In addition there are some interdisciplinary primary journals such as *Nature, Science, Experientia*. These are general research publications that cover most of the branches of science with emphasis on the biological and/or physical sciences.

One of the major problems in scientific publications is the long time required for publication. It usually takes 6–8 months (sometimes even more) for the publication of a full paper or a note and 3-4 months for the publication of a communication. The reason for such a long time is the complex mechanism involved in the transformation of the typed manuscript into a printed article. One has to remember that the communication channel between the author and the editor, the referees and the editor as well as between the printer and the author is still the written word via the national or the international mail service.

2.3. Trends in scientific publications

In addition to the long publication timelag the scientific publications suffer from ever-increasing production costs; these increased costs are reflected in a higher subscription rate. Many journals today have a subscription rate above 500 dollars and it is not rare to find journals with an annual subscription rate exceeding 1000 dollars. Some typical 1986 subscription rates are given in Table 2.5. A survey that was carried out in 1977 by F.F. Clasquin and J.B. Cohen [*Science*, **197**, 432 (1977)] showed that the average price of the chemistry journals in the period 1967–1976 increased at a rate 75% greater than the consumer price index in the USA (the figure for physics journals was 100%). There are good reasons to believe that the situation in the last decade has become even worse. The rising costs of the scientific journals has already forced some libraries to reduce drastically, or even eliminate entirely, the purchase of books, in order to maintain their journal and serial subscription. Even so, many libraries, and their number will increase during the coming years, have had to cut some of their subscriptions. Thus increasing costs, decreasing circulations and long publication timelag reinforce the arguments, made on information flow consideration, that the primary research journal needs some fundamental change.

Table 2.5   Some 1986 journal subscription rates
(in US dollars)

Chemische Berichte	478
Journal of Molecular Catalysis	605
Journal of Neurochemistry	645
Journal of Non-Crystalline Solids	1048
Journal of Materials Science	1115
Journal of Molecular Biology	1250
Journal of Organometallic Chemistry	1953
Biochimica Biophysica Acta	2925

How could the scientific community change and simplify the publication mechanism in order to decrease production time and cost which will be followed by increased circulation?

2.3.1. Direct reproduction journals

The simplest approach to reducing the timelag between acceptance date and publication date is to reproduce the original manuscript directly by photography instead of conventional composition. This enables the publisher to eliminate the technical editing of the manuscript, the composition, the proofing, and the correction stages. All this results in large savings of time and money. Although the first journals to use this approach were the rapid communication journals (e.g. *Biophysical and Biochemical Research Communications*, *Tetrahedron Letters*), there are today regular journals (e.g. *Journal of Chromatographic Science*, *Life Sciences*) and conference proceedings (e.g. *Proceedings of the 6th International Conference on the Origin of Life and the 3rd International Meeting of ISSOL*, published by D. Reidel, 1981) being published by direct photography of the authors' transcripts.

2.3.2. The two package concept

Another approach is the two package concept, such as the dual journal. The dual journal approach is based on a two-part publication which is issued simultaneously. The first part is a printed journal of summaries or synopses, each of which are accompanied by necessary charts, graphs, tables and leading references and do not exceed one or two printed pages. These summaries or synopses give the chemist a general account of the scientific discovery and an idea about the methods used. The second part is a reproduction in the form of microfilm, microfiche and/or miniprint of the full, original manuscript. Individuals usually would subscribe only to the synopsis journal while libraries would order the dual package—the synopsis journal plus the microform version (in case there is more than one microform version available they may order more than one version of the latter part of the package).

The best known journal of this kind is the *Journal of Chemical Research*, which is published jointly by the British, French and German Chemical Societies. It appears as a synopsis journal plus a microfiche and a miniprint version.

The dual journal concept not only helps us to cope with the rising cost of scientific publications and to reduce publication time, but also to cope with another problem in scientific communication, that of browsing. A journal reader scans a large amount of unnecessary information, as much more material is provided to him than is needed or can be used. Few, if any, of the papers published in a journal are read through from the beginning to the end by more than a tiny fraction of the readers. Further, the chemist today finds too many papers of no interest to him at all in each issue of the journals to which he subscribes (or reads), while papers of interest to him are spread over many journals that he does not read at all.

These problems could be solved by presentation and packaging of the scientific information in different ways. In the dual journal approach, the synopsis journal gives the reader relevant information from the standpoint of the main thrust of the work and its conclusions. The specific detailed information, which only a few of the readers need, is reported in the microform or miniprint journal. Thus those interested in this information could locate it there. Another example of two different packages is the supplements to papers published in some of the journals of the Royal Society of Chemistry and the American Chemical Society. These supplements contain detailed specialized information (such as X-ray data, computer programs, spectroscopic data, and analytical data) which are of interest to only a few workers in the field. The supplementary material is seen and evaluated by the referees, but is subsequently sorted separately and is available in a microform or a photocopy from the publisher for a moderate price. A few years ago the editorial board of the *Journal of the American Chemical Society* started recommending and encouraging authors of communications to submit the experimental details of their work as supplementary material.

A more extreme approach to the problem can be found in the USSR where deposition of manuscripts is taking place. The manuscripts are deposited either directly with the All Union Institute for Scientific and Technical Information (VINITI) or with one of the many All Union deposit centres in the USSR. Titles of all such deposited works are published in the *Catalog of Deposited Manuscripts* and are abstracted in the *Abstract Journal of VINITI* (*Referativnyi Zhurnal*). Deposition can take place also via a primary journal (in which case an abstract or just a title and a reference number are published in the journal concerned).

2.4. The journal of the future

Many people wonder about the future of the scientific journals. While the costs of paper, postage, labour, etc. are going up the price of computer storage, online access, data communication, microfiche, audio/visual casettes and other paperless means of communication is going down. Online versions of some of the scientific journals are available to the public (e.g. most of the journals published by the American Chemical Society—*Accounts of Chemical Research, Chemical Reviews, Inorganic Chemistry, Journal of the American Chemical Society* etc.; most of the journals published by the Royal Society of Chemistry—*Journal of the Chemical Society Dalton Transactions, Journal of the Chemical Society Perkin Transations I* and *II, Journal of the Chemical Society Chemical Communications, Journal of Chemical Research Synopses, The Analyst* etc.; Wiley polymer journals—*Biopolymers, Journal of Applied Polymer Science, Journal of Polymer Science Part A, Part B* and

Part C; as well as *Angewandte Chemie International Edition*, are available via STN). Very recently electronic journals started to become available. Does all this mean that the paperless revolution will be the cause of the disappearance of the hard-copy journal? The answer is no!

The hard-copy journals will exist in the future because there is a need for them! The scientist must receive some kind of feedback and, in order to get it, he has to have a broad scientific view in general and in his own field in particular. A scientist must be able to browse through the current literature. A researcher must be able to see and feel on a continuing basis the whole thrust of where his discipline is going. This is important, as from the past we have learnt that most, if not all, first class ideas and brilliant scientific solutions to various problems, come from people who developed their expertise in one field and suddenly saw the opportunity to apply it to a totally different field, or from people who saw the opportunity of applying a novel technique developed in a different area by other workers to solving the problems in their own area. The journals of the future will be different from the present ones, but scientific journals have always changed over the years, and journals of the mid 1980s cannot be compared to those of the early 1960s or late 1950s.

Looking at journals of the future one has to take into consideration the rising costs of subscriptions and tight library budgets. The situation will become worse in the future, when more and more libraries will have to reduce the number of journals to which they subscribe. Publishers will not be able to depend on a certain circulation to libraries of a few thousand copies for any old or new scientific journal. The assumption that libraries are and will continue to be buying everything is not true any more, and those libraries that are still doing so will be forced to change their policy in the near future. Marginal quality journals will disappear as libraries will be unable to purchase them, therefore making publishers unable to publish them.

Scientists will continue to do research and write accounts of their work. With the disappearance of many of the traditional scientific journals a large number of papers will be deposited in deposit centres, and the scientific community will be able to retrieve them by online access or by the use of secondary information services. There will be a switch to more cost-effective means of publication. Electronic journals (journals in which interaction between editors, authors, reviewers and/or readers is carried out by means of computer communication) will become available to the general public. A large number of journals will adapt the dual journal concept in order to survive. Others will look for new formats and ways of presentation to provide the scientific community with the information it needs, keeping the price tag at a reasonable level. For example, the distribution of complete copies only of those papers which are of specific interest to the particular subscriber is an idea which has been discussed in the last few years. Another idea is to have a communication-type journal where the detailed experimental data will be available to interested readers by machine-retrievable means.

To summarize, the hard copy journals are here to stay, but their format and style will change, and their number will decrease. Probably only a few major prestige multi-disciplinary journals will dominate, with a core of a small number of extremely specialized journals of the highest quality. These first-class hard-copy journals will be supplemented by specially selected and distributed papers, abstracts and titles to meet specific interests and requirements.

CHAPTER 3

The Library

All published material—journals, reports, abstracts, monographs, books, theses, dictionaries, directories etc.—is to be found in the library. As one will probably spend considerable amounts of time in the library, one should do two things: (a) learn about the structure and the organization of the library in general and of one's own organization's science or chemistry library in particular, and (b) establish a good relationship with the librarians (or the information specialists)—they are always highly trained people, knowledgeable in the area of information sources, and are ready to assist.

3.1. Organization of the library

3.1.1. Library classification systems

How should we organize a library to satisfy everyone's needs? The simplest way may be to arrange the various books according to their size; aesthetically it might be nicer to arrange the books according to the colours of their covers; another easy way is to put the books in alphabetical order of the authors. These and similar arrangements have the disadvantage that books dealing with the same subject, or even two volumes of the same series, would end up on two different shelves or even in two different corners of the library. Some kind of classification which would place books that are alike together should be used. As most science libraries are open-stack libraries, knowing the classification method used by one's library would enable the user not only to find the specific volume he is looking for easily, but also to find all the other material on the same subject held by the library.

Establishing a classification system is not simple. A book can be placed in one, and only one, place in the library. Books dealing with two (or more) subjects (or division of a subject) are placed in the subject or division which appears to be the most important one. Thus C.L. Perrin's book *Mathematics for Chemists* could be classified under mathematics, if we assume that this is the most important subject, or under chemistry, if the assumption is that the emphasis is on its users—the chemists. Indeed, one classification system classified the above book among mathematics textbooks, while another system classified it under problems and exercises in the chemistry division.

The situation is different when the various subjects or divisions are of equal importance. In such cases the book is classified in the subject treated first; e.g. H.R. Allcock's book *Phosphorus-Nitrogen Compounds* is classified under inorganic phosphorus compounds.

Many library classification systems exist; some are widely used whereas others are used by only very few libraries (some-times by only one or two libraries). We shall discuss here briefly the three main classification systems: The Dewey Decimal Classification, The Universal Decimal Classification and The Library of Congress Classification.

3.1.1.1. The Dewey Decimal Classification

The Dewey Decimal Classification (DDC) system is a general classification system devised by Melvin Dewey during the 1870s. Dewey divided the whole field of knowledge into nine major classes (number 100 to 900), with a tenth class devoted to general works (numbered 000) (Table 3.1). Each main class is subdivided into ten divisions, e.g. class 500—Pure science division—is given in Table 3.2.

Table 3.1   The 10 main classes of DDC

000	Generalities
100	Philosophy and related disciplines
200	Religion
300	Social sciences
400	Language
500	Pure sciences
600	Technology (applied sciences)
700	The arts. Fine and decorative arts
800	Literature
900	General geography and history and their auxiliaries

Table 3.2   The 100 divisions of pure science in DDC

510	Mathematics
520	Astronomy and allied sciences
530	Physics
540	Chemistry and allied sciences
550	Science of the earth and other worlds
560	Palaeontology. Palaeozoology
570	Life sciences
580	Botanical sciences
590	Zoological sciences

Each of the divisions is subdivided again into ten subdivisions, e.g. class 540, which is the chemistry division, is subdivided as described in Table 3.3.

Table 3.3   The 1000 divisions of chemistry and
allied sciences in DDC

541	Physical and theoretical chemistry
542	Chemical laboratories, apparatus, equipment
543	Analytical chemistry
544	Qualitative chemistry
545	Quantitative chemistry
546	Inorganic chemistry
547	Organic chemistry
548	Crystallography
549	Mineralogy

Table 3.4   Subdivision of inorganic chemistry in DDC

546		Inorganic chemistry
		Chemistry of elements of inorganic compounds and mixtures. Class here physical and theoretical chemistry, analytical chemistry, physics of specific elements, compounds, mixtures, groupings; comprehensive work on inorganic and organic chemistry of specific elements, compounds, mixtures, groupings
		*
		*
546.38		Alkali metals (group 1A)
546.381	Li	Lithium
	1	The element
	2	Compounds
	22	Acids and bases
	24	Salts
	25	Complex compounds
	3	Molecular
	4	Theoretical chemistry
	5	Physical chemistry
	6	Analytical chemistry
	64	Qualitative analysis
	65	Quantitative analysis
546.382	Na	Sodium
		*
		*
546.39		Alkali earth metals (group 2A)
		*
		*
547.75		Rare gases (group 8A)
		*
		*
547.756	Rn	Radon

Further divisions are made using decimal numbers. The further division of
546—Inorganic chemistry—is given in Table 3.4. The detailed classification
can be found in the latest edition of the *Dewey Decimal Classification and
Relative Index* which is published by Forest Press (Albany, NY 12206, USA).

3.1.1.2. The Universal Decimal Classification

The Universal Decimal Classification (UDC) system was originally a European adaptation of the DDC. It followed the same general rules as the DDC except that a main class is designated by one digit (from 0 to 9), a subclass or a division is designated by two digits, and subdivision is designated by three digits (see Tables 3.5–3.7). Any further classification is made by adding digits. Long numbers are broken into three-digit groups by inserting decimal points. During the years the International Institute of Bibliography and its successor, the International Federation for Documentation (FID) (7 Hofweg, 2511 AA, The Hague, The Netherlands), realized the need to change the assigned classification numbers owing to the development of new areas of knowledge. While the DDC adopted the attitude that numbers once assigned and printed should not be changed (an attitude which was softened slightly during the 1950s), the UDC was ready for adaptation according to changes and need. If a UDC number was found to be old or non-usable, this number could be cancelled and a new free number could be given to the same subject. Cancelled numbers become free ten years after the date of cancellation and can be reused. This period is needed in order to enable the various libraries using the classification system to adapt themselves to the change. Those who are interested in the whole classification scheme should use the complete *English Tables* which were published over the years, or the *Abridged English Tables* which were published in 1961 (FID publication No. 289) and the ten-year supplement to the abridged UDC editions: 1968–1969. (Editions in other languages are also available.) An excellent introduction to the system is the *Guide to the Universal Decimal Classification* published by the British Standards Institution, London, 1963. More information can be found in H. Wellisch's book *Universal Decimal Classification, A Programmed Instruction Course* as well as in A.C. Foskett's book *The Universal Decimal Classification; The History, Present Status, and Future Prospects of a Large General Classification Scheme*. The FID also publishes *Extensions and Corrections* to the UDC. These are semiannual publications describing the various changes and additions to the classification system, keeping the published edition up to date.

The DDC and the UDC used to be remarkably similar in their main structure, although differences in a third digit could be found (e.g. the number for Biochemistry is 574.19 in the DDC system and 577.1 in the UDC system). During the late 1950s some changes occurred in the 544 and 545 subdivision of the UDC. Radical changes in the UDC took place in the early 1960s all of the material in class 4 (Linguistics and Philology) being transferred to class 8 (Literature). At present class 4 remains vacant and no decision has yet been taken on its future use. There is also an active international committee working at present on improving the tables for chemistry (54—chemistry, crystallography, mineralogy; 66—chemical and related industries, chemical technology).

Table 3.5   The 10 main classes of UDC

0	Generalities of knowledge
1	Philosophy, Metaphysics, Psychology, Logic, Ethics
2	Religion, Theology
3	Social sciences
4	—
5	Mathematics and natural sciences
6	Applied sciences, Medicine, Technology
7	The Arts, Recreation, Entertainment, Sport
8	Languages, Linguistics, Literature
9	Geography, Biography, History

Table 3.6   The 100 divisions of mathematics and natural sciences in UDC

51	Mathematics
52	Astronomy, Astrophysics, Space Research, Geodesy
53	Physics
54	Chemistry, Crystallography, Mineralogy
55	Geology, Meterology, Hydrology
56	Palaeontology
57	Biological sciences
58	Botany
59	Zoology

Table 3.7   The 1000 divisions of chemistry, crystallography, mineralogy in UDC

541	General, theoretical and physical chemistry
542	Experimental, preparative chemistry, laboratory equipment
543	Analytical chemistry
544	—
545	—
546	Inorganic chemistry
547	Organic chemistry
548	Crystallography
549	Mineralogy

3.1.1.3. *The Library of Congress Classification*

The Library of Congress (LC) Classification system is not based on any philosophical or theoretical grounds. It is a practical system which was developed in order to classify the collections of the Library of Congress. Towards the end of the 19th century reorganization was taking place at the Library of Congress and, as part of this reorganization, the library was trying to adapt a new classification system which would enable it to classify all of its present and future collections. As a result of studying all of the existing classification systems at that time, the system most suitable for adaptation was found to be the DDC. A major difficulty in adapting the system was Dewey's refusal to allow any change to it. The result was that the Library of Congress designed a new classification system for its own use. No attempts were made to create a general, perfect classification system which could be used by other libraries (even so, a large number of libraries adapted the LC Classification system and are using it today).

In the LC system all books are divided into 21 main classes, which are designated by one letter, as can be seen in Table 3.8.

Table 3.8   The main classes of LC classification

A	General works
B	Philosophy, Psychology, Religion
C	Auxiliary sciences of history
D	General and old world history
E,F	American history
G	Geography, Anthropology, Recreation
H	Social sciences
J	Political sciences
K	Law
L	Education
M	Music, Books on music
N	Fine arts
P	Philology, Linguistics, Literature
Q	Science
R	Medicine
S	Agriculture
T	Technology
U	Military sciences
V	Naval sciences
Z	Bibliography, Library science

Classes are divided into subclasses which are designated by two letters (e.g. the subdivision of Science which is designated by Q is shown in Table 3.9).

Table 3.9.   Science subclasses in LC classification

QA	Mathematics
QB	Astronomy
QC	Physics
QD	Chemistry
QE	Geology
QH	Natural history (general), Biology (general)
QK	Botany
QL	Zoology
QM	Human anatomy
QP	Physiology
QR	Microbiology

The division of the subclasses are designated by integral numbers in ordinary sequence ranging from 1 to 9999 which may be extended decimally.

Further subdivision is possible by using a combination of initial letters followed by Arabic numbers; e.g. QD 181 is a chemistry subdivision dealing with special elements and differentiation between different elements is done by using further subdivision—N1 designated nitrogen, O1 designated oxygen, P1 designated phosphorus, etc. (see Table 3.10).

Table 3.10   Subdivision of subclass in LC classification

QD	181 Special Elements—arranged alphabetically according to the element symbol. In this subdivision are included works on the origin, properties, reactions and analytical chemistry of the elements and their inorganic compounds (excluding the determination of their atomic and molecular weights).

A1 = A	Argon	B7 = Br	Bromine
A2 = Ac	Actinium	C1 = C	Carbon
A3 = Ag	Silver	C15= CN	Cyanogen
A4 = Al	Aluminium	C2 = Ca	Calcium
A5 = Am	Americium	C3 = Cd	Cadmium
A7 = As	Arsenic	C4 = Ce	Cerium
A8 = At	Astatine	C5 = Cl	Chlorine
A9 = Au	Gold	C55= Cm	Curium
B1 = B	Boron	C6 = Co	Cobalt
B2 = Ba	Barium	*	
B3 = Be	Beryllium	*	

A detailed classification of the various sections can be found in the *Library of Congress Classification* which is issued in 32 volumes. Each volume is devoted to a class (section) or a part of a class. It is revised from time to time and reissued separately.

It is interesting to mention here the classification system of the National Library of Medicine (Washington DC, USA). This classification covers the whole field of medicine and its related sciences. The notation was developed from the block of letters QS–QZ and W, which are unused in the LC Classification system.

All classification systems give a classification number to the book. Usually there are few books having the same classification number. Each book has its own call number which is composed of its classification number followed by an initial letter and an Arabic number indicating the author (usually known as the author number) and by any other numbers that will indicate edition(s) or volume number.

3.1.2. The catalogue

The catalogue is the record of the books in the library, giving the reader the general information which is found on the title page of the book. One can find the following items listed: call number; author's name, book title, edition number (if it is not the first edition of the book), translator's name (if the book has been translated from another language), publication data (e.g. publisher, date of publication), book information (e.g. number of volumes and/or pages, data concerning maps, tables etc.), serial number (ISSN) (if the book is part of a serial publication) or book number (ISBN), special features (e.g. bibliography), and table of contents (whenever the book contains several works by one or several authors).

Most libraries today are computerized libraries, thus they are using computers for their housekeeping operation—acquisitions, cataloguing, circulation, serial control etc. The computer is used to produce and maintain the catalogue which is available to the user in a machine-readable form (using a cathode ray tube (CRT), or a computer console). The old card catalogue remains in many libraries beside the computerized catalogue; other libraries have discarded the card catalogue completely. The Library of Congress card catalogue was frozen on 31 December 1980. Subsequent acquisitions are listed in (and only in) the computerized catalogue.

Working with the card catalogue one has to keep in mind that there is more than one card per book. In addition to the main entry card there are further cards in which the necessary heading is added above the author's name. These added entry cards are: title card, subject cards (usually more than one, information about subject headings can be found in the latest edition of *Library of Congress Subject Headings*), joint author cards (a card for each joint author), and a serial card (if the book is part of a series).

All the cards are filed in the card catalogue, which is usually a dictionary catalogue, i.e. all of its entries (author, subject, title etc.) are arranged together in one alphabetical order.

There are libraries that classify and catalogue journals in the same way as books, so that journals and books in the same field are placed together. Others catalogue journals separately, and place them together alphabetically by title in a separate area. Some libraries even separate the journals themselves between recent volumes (the last 10–15 years) and old volumes, placing the old periodical volumes in the basement or in another less accessible location.

3.2. Locating material in the library

3.2.1. Locating a book

How can one find a book in the library? The answer is simple: one should find the book call number. How can one find the specific book call number? The problem is very simple once the bibliographic data of the desired book is known, especially the author(s) name(s) and the book title. In order to try to locate F.H. Portugal and J.S. Cohen's book, *A Century of DNA, A History of the Discovery of the Structure and Function of the Genetic Substance*, all that one has to do is to look under Portugal, F.H. in the library catalogue and find the catalogue card (using the card catalogue) (Fig. 3.1) or the corresponding screen (using the computerized catalogue) (Fig. 3.2) (a menu screen of a computerized catalogue is shown in Fig. 3.3). The catalogue card or the corresponding screen will give us the book call number (in the LC Classification system the call number is QP624P67.) The same is true when only the title of the book is known. Looking under 'Century of DNA' in the catalogue would reveal a title entering card or a display screen with the identical information.

> **Portugal, Franklin H**
>   A century of DNA : a history of the discovery of the structure and function of the genetic substance / Franklin H. Portugal and Jack S. Cohen. — Cambridge, Mass. : MIT Press, c1977.
>   xiii, 384 p. : ill. ; 24 cm.
>   Includes bibliographical references and index.
>   ISBN 0-262-16067-6
>   1. Deoxyribonucleic acid—History.   2. Molecular genetics—History.   3. Molecular biologists—Biography.   I. Cohen, Jack S., joint author.   II. Title.
>   QP624.P67   574.8'732   77-7340
>   77   MARC

Figure 3.1.   Library catalog's book card.

```
BIBLIOGRAPHIC RECORD 0896548
CALL. NO   QP 624 P67
  AUTHOR   PORTUGAL, FRANKLIN H.
AUTHOR-A   COHEN, JACK S.===JOINT AUTHOR
   TITLE   A CENTURY OF DNA
  SUB-TL   A HISTORY OF THE DISCOVERY OF THE STRUCTURE
           AND FUNCTION OF THE GENETIC SUBSTANCE
   PLACE   CAMBRIDGE, MASS.
 PUBLISH   MIT PRESS
    YEAR   1977, REPRINTED 1978
   PAGES   XIII, 384 P.
 SUBJECT   DEOXYRIBONUCLEIC ACID - HISTORY
           MOLECULAR GENETICS - HISTORY
           MOLECULAR BIOLOGISTS - BIBLIOGRAPHY
     LAG   ENG
METR-TYP   M
  YEAR-F   1977
    ISBN   0262160676
 LC-CARD   77007340
 TITLE-C   FRANKLIN H. PORTUGAL AND JACK S. COHEN
```

Figure 3.2.   Computerized library catalog's book screen.

The situation is more complicated if one is looking for a book on a specific subject without any bibliographic information. How can a book on the history of DNA research be located? The simplest way would be to look for a subject entry card or screen—to look for the suitable subject heading. The most reasonable subject heading should be desoxynucleic acid. Indeed,

```
C A T A L O G     S E A R C H

To search by code type:    CODE/TERM
          then press        ENTER

TO SEARCH BY -                      CODE

Title.............................TL
Word from the title..............WTL
Author,editor, etc...............AU
Word from name of sponsoring body
     or author...................WAU
Subject heading..................SH
Word from subject heading........WSH
Call number......................CN
Location, copies, status.........H
Description of item..............B

          Return to menu  SE
```

Figure 3.3.   Computerized library catalog's menu screen.

using desoxynucleic acid as a subject heading results in locating a subject entry card or a screen giving full information on F.H. Portugal and J.S. Cohen's book, *A Century of DNA, A History of the Discovery of the Structure and Function of the Genetic Substance.*

It may occur that no book can be found by using subject entry words. This does not necessarily mean that the desired book is not present in the library, as it could be that the wrong subject entries were used. In order to get a better definition of the subject one should always consult the latest edition of *Library of Congress Subject Heading.* If this does not work, the last resort is to determine the class number of the desired book according to its subject matter and to locate all the books that have this class number using either the classified catalogue (a catalogue arranged by class number rather than in alphabetical order), the shelf list (records of books in the library arranged according to their order on the shelves), or the classification number field in the computerized catalogue (a field which is arranged by class numbers).

The determination of a class number for a book is relatively a simple procedure. Details are somewhat different in different classification systems. However, in all cases the corresponding system tables are used; e.g., in libraries using the DDC one uses the latest edition of *Dewey Decimal Classification and Relative Index* and its *Schedule*, in libraries using the LC Classification, the latest edition of the corresponding volume of *Library of Congress Classification* is used.

However, one has to keep in mind that there are cases in which books which are alike are classified differently. Those are rare cases but one has

24

to be aware of it whenever one is looking for a book class number. Two examples are given below:

a. 1. J.H.D. Eland, *Photoelectron Spectroscopy: An Introduction to Ultraviolet Photoelectron Spectroscopy in the Gas Phase.*
2. J.W. Rabalais, *Principles of Ultraviolet Photoelectron Spectroscopy.*
While the Dewey class number of the first book is 543.0858 (analytical chemistry) the Dewey class number of the second book is 539.1 (atomic physics). The two books are introductory books to UV photoelectron spectroscopy.

b. 1. D. Swern, Ed., *Organic Peroxides.*
2. E.G.E. Hawkins, *Organic Peroxides: Their Formation and Reaction.*
3. A.G. Davies, *Organic Peroxides.*
While the LC class number of the first book is QD305E7 (organic chemistry: aliphatic compounds, ethers and oxides), the LC class number of the second book is QD305H7 (organic chemistry: aliphatic compounds, alkenes), and the third book LC class number is 412.01 (organometallic compounds, oxygen compounds). The three books are devoted to the formation and reaction of organic peroxides.

3.2.2. Locating an article in a journal

Locating a specific scientific article is a simple task once the full bibliographic information of the article one is looking for (or the reference) is known. In order to locate Hans Kuhn's paper on the evolution of the genetic code, which was published in 1972 on page 798 in Volume 11 of *Angewandte Chemie, International Edition* [H. Khun, *Angew. Chem., Int. Ed. Engl.*, **11**, 798 (1972)], all that one has to do is to locate the desired journal (in this case *Angewandte Chemie, International Edition*) place in the library, pick up the desired volume (Volume 11), and open it at the right page (page 798).

3.2.2.1. References

References or bibliographic citations are given in a scientific work (e.g. journal articles, reports, grant proposals, theses, books) not only as a direction to the source of the information reported but also as a credit to fellow scientists for their work. They are given in an accepted abbreviated form, but this may have a different style in different publications.

The authors' names and initials are given first, followed by the journal title. The latter is given in abbreviated form [the list of approved journal abbreviations according to the International Union of Pure and Applied Chemistry (IUPAC) is given in the *Chemical Abstract Source Index (CASSI)* or in the latest edition of *International List of Periodical Title Word Abbreviations* published by Biblio Verlag], followed by the volume number (which is usually given in bold type or underlined), the page number(s) and the year; the order of these details may vary. Issue numbers are usually given only when citing a journal whose pages are numbered separately in each issue (e.g. *Chemical and Engineering News*).

Thus the cited reference to Kuhn's paper could be given in various forms:

H. Kuhn, *Angew. Chem., Int. Ed. Engl.*, **11**, 798 (1972).
H. Kuhn, *Angew. Chem., Int. Ed. Engl.*, (1972), **11**, 798.
Kuhn, H., *Angew. Chem., Int. Ed. Engl.*, **11**, 798–820 (1972).
Kuhn, H., (1972), *Angew. Chem., Int. Ed. Engl.*, **11**, 798–820.

When one submits a paper for publication in a journal, one should write the references according to the particular house style. The situation is different when one is arranging a list of publications in internal reports or grant applications. In this case one can use whichever style is preferred, but that style should be used consistently.

Some journals use the words *This Journal* when citing references to work published in it previously, and others use the abbreviation *ibid.* when referring to work published in a journal already cited in the previous reference.

The style of citing a book is different. The authors' names and initials are given first, followed by the title of the book, the Editor's name (if applicable) and the edition number (if it is not the first edition), then the name of the publisher, the place of publication, the year of publication and the relevant page numbers.

3.2.2.2. *Using journal indexes*

How can one locate a desired paper when only part of the bibliographic data are available? For example, how is the T.C. Ehlert paper on the thermochemistry of copper fluorides which was published in the *Journal of Physical Chemistry* in 1977 to be found? One way is to go through the 1977 volume(s) of the journal page by page in order to locate the paper. Only one volume of the *Journal of Physical Chemistry* was published in 1977—Volume 81 totalling 2683 pages. However, this solution is impractical as going through 2683 pages would take a long time. All journals (with few exceptions) publish an Author and a Subject (or Keywords) Index at the end of each volume or at the end of the year. Knowing the author's name, the desired paper can be located very easily. Examining the Author Index of Volume 81 of the *Journal of Physical Chemistry*, the desired entry is found under Ehlert, T.C., 'Thermochemistry of Copper Fluorides', 2069. All one has to do now is to turn to page 2069. After locating the article one should write the full reference in one's personal files for further use [e.g. in this case—T.C. Ehlert and J.S. Wang, *J. Phys. Chem.*, **81**, 2069 (1977)].

The situation is more difficult when one cannot recall the author's name, e.g. one may remember that some time in 1977 or 1978 one saw a paper in *Proceedings of the National Academy of Science of the USA*, describing the use of enzymes as reagents in organic synthesis, possibly connected with peptide synthesis. Checking the Subject Indexes of the above journal for the years 1977 and 1978, using the terms enzyme, synthesis, organic synthesis, enzymatic synthesis, bioorganic synthesis, reagents, chemical reagents, and bioorganic reagents gave negative results. However under the term peptide synthesis in the 1977 Subject Index one could locate the following entry:

Peptide Synthesis
    Enzymes as reagents in peptide synthesis
    Enzyme-labile protection for carbonyl group—2739

Again the full reference should be noted for future use: J. Glass and M. Pelzig, *Proc. Natl. Acad. Sci. USA*, **74**, 2739 (1977).

3.2.2.3. *Using* Chemical Abstracts

Everything is relatively simple when we know the journal name and the publication period, but what does one do when one has no idea in which journal and/or when the desired work was published? Sometimes one even does not know who carried out the work, but only recollects seeing somewhere a paper on the subject. In all such cases one has to turn to *Chemical Abstracts* (*CA*) for help.

CA publishes concise informative abstracts of journal articles, published conference papers, technical reports, deposit documents, dissertations, and patents, as well as book announcements in chemistry. All abstracts and announcements contain the complete bibliographical information needed to locate the original document. Since 1962 the *CA* appears weekly; the odd-numbered issues are devoted to biochemistry and organic chemistry and the even-numbered issues are devoted to macromolecular chemistry, applied chemistry and chemical engineering, and physical and analytical chemistry. At the end of each issue there are three indexes: Author Index, Keyword Index and Patent Index.

Since 1962 two volumes of *CA* have been published annually. Prior to 1962 one volume covered a whole year. The volume indexes are cumulated every ten volumes (at present every five years; prior to 1962 every ten years) into Collective Indexes (e.g. the Tenth Collective Index covers the years 1977–81 (Volumes 86–95), the Fifth Collective Index covers the years 1947–56 (Volumes 41–50)). The Sixth Collective Index is an exception, covering only five volumes—Volumes 51–55, which were published in the years 1957–61.

CA is a dynamic system which has been changed over the years. These changes affected among other things its indexing policies. The present index system contains the following Volume and Collective Indexes: Author

Index, Chemical Substance Index, General Subject Index, Formula Index, and Patent Index. Prior to 1972 a single Subject Index was published; it includes chemical compounds as well as general subjects arranged in a single alphabetical order. Prior to 1977 two Patent Indexes were published: the Numerical Patent Index and the Patent Concordance. Table 3.11 describes the various individual indexes which comprise each of the Collective Indexes.

An excellent explanation and illustration of the organization and use of the various *CA* indexes is given in the *Index Guide*, which also gives an extensive summary of the *CA* nomenclature, and (starting with Volume 85) information about the hierarchical order of the General Subject Index.

The Index Guide provides cross-references from general subject terms and chemical substance names used in the scientific literature and/or prior indexes of *CA* to the terminology used in the current indexes. It used to be published at the begining of each five year collective period followed by a yearly supplement, which gave new information from the previous supplements. Since the begining of the 11th Collective Index period no supplements to the *Index Guide* are published. Instead a complete update version of the *Index Guide* is published about every 18 months.

How can one locate what Felix Bergmann published in 1978 on enkephatlin?

One uses the *CA* Author Index. In order to retrieve Felix Bergmann's works on enkephalin which were published in 1978 one has to use the *CA* 10th Collective Author Index which covers the years 1977–1981. The Author Index lists personal and corporate authors, including patentees and patent assignees, with the title of their publications and the corresponding abstract number. There is a cross-reference from secondary authors to principle first-named authors. (Prior to 1967 *CA* abstracts did not have numbers and the various *CA* indexes directed users to the page where the corresponding abstract was published.)

In the 10th Collective Author Index there are fourteen entries under Bergmann, Felix and two entries under Bergmann, F. (Fig. 3.4) (as an author name may appear in more than one form and the exact form(s) are not always predicted, it is recommended to examine the various name forms). One of the sixteen entries (90: 145718h) seems to be the desired one (although Volume 90 of the *CA* was published in the first half of 1979, one has to keep in mind that there is a delay of a few months from the publication date of an article in a journal until the work is abstracted in the *CA* — indeed the above work was published in 1978). There are few cross-references to other papers which Felix Bergmann (F. Bergmann) is a coauthor: Diller, Dov; Feret, Peter P.; Tanner, Stephen J.; Von Voss, Hubertus; Weissman, Ben Avi. All of those index entries should be checked for additional relevant references. Doing so one finds additional relevant references under Weissman, Ben Avi (90: 180078w).

Table 3.11 CA collective indexes

Collective index	1st	2nd	3rd	4th	5th	6th	7th	8th	9th	10th	11th	12th
Years	1907–16	1917–26	1927–36	1937–46	1947–56	1957–61	1962–66	1967–71	1972–76	1977–81	1982–86	1987–91
CA volumes	1–10	11–20	21–30	31–40	41–50	51–55	56–65	66–75	76–85	86–95	96–105	106–115
Author index	X	X	X	X	X	X	X	X	X	X	X	X
Subject index	X	X	X	X	X	X	X	X				
Index guide								X	X	X	X	X
Chemical substance index									X	X	X	X
General subject index									X	X	X	X
Formula index		*	*	*	X	X	X	X	X	X	X	X
Numerical patent index					X	X	X	X	X			
Patent concordance							X	X	X			
Patent index										X	X	X

*There is a Twenty-Seven Year Collective Formula Index to CA (1920–1946) covering Volumes 14–40.

——; McLellan, E. J.; Webb, J. A.
Fast nondamageable laser–pulse detector using
   gaseous plasma, 93: 104672g
Bergmann, F. See Von Voss, Hubertus
——; Elam, R.
On the mechanism of action of 2–amino–4–=
   methylpyridine, a morphine–like analgesic, 94:
   58196a
——; Feldberg, W.
Effects of propylbenzilylcholine mustard on injection
   into the liquor space of cats, 89: 157291f
Bergmann, Felix See Diller, Dov; Diller, Dov.; Feret,
   Peter P.; Tanner, Stephen J.; Weissman, Ben Avi
——; Altstetter, R.; Pasternak, V.; Chaimovitz, M.;
   Oreg, M.; Roth, D.; Hexter, C. S.; Wilchek, M.
Cerebral application of enkephalins, 86: 115537y
——; Altstetter, R.; Weissman, Ben Avi
In vivo interaction of morphine and endogenous
   opiate–like peptides, 90: 145718h   ◄————
——; Brown, D. J.
Methylation of purin–8–ones, 91: 123066n
——; Diller, Dov
Fragmentation of a new type of 6–phenyl–2–pyridone

Figure 3.4   A *CA* Author Index

The answer to the question can now be given on the bibliographic information reported in abstracts 90 : 145718h and 90 : 180078w. Felix Bergmann published in 1978 two papers dealing with enkephalin:

F. Bergmann, R. Altstetter and B.A. Weissman, *Life Sci.*, **23**, 2601–8 (1978)
B.A. Weissman, F. Bergmann and R. Altstetter, *Dev. Neurosci.* (Amsterdam), **4**, 297–8 (1978).

3.3.   Online search

The online searching (or computerized literature searching) is one of the 'spinoffs' of the computerization of the various information services. Following the computerized photocomposition revolution in the printing industry in the early 1960s, the various information services started to computerize the editing and indexing processes of their abstracts. The net results of these changes were that besides the printed products the information service products became available in a computer readable form and could be searched using a suitable software.

Online search is carried out by communication with a computer search system. The communication is carried out via a computer terminal (a teletype or a cathode ray tube—CRT) which is linked to the computer through a telecommunication system (generally via ordinary telephone lines).

We are not going to discuss in detail the various aspects, strategies and principles of online searching. There is a large number of books devoted to this subject, some of which are:

W.M. Henry, J.A. Leigh , L. A. Tedd and P.W. Williams, *Online Searching: Introduction*, Butterworths, London, 1980, 209pp.

C.C. Chen, and S. Schweizer, *Online Bibliographic Searching: A Learning Manual*, Neal-Schuman Publishers, New York, 1981, 227pp.

C.H. Fenichel, and T.H. Hogan, *Online Searching: A Primer*, Learned Information, Marlton, N.J., 1981, 152pp.

E.P. Hartner, *Introduction to Automated Literature Searching*, Marcel Dekker, New York, 1981, 145pp.

C.T. Meadow and P.A. Cochrane, *Basics of Online Searching*, Wiley-Interscience, New York, 1981, 244pp.

B. Gerrie, *Online Information Systems: Use and Operating Characteristics, Limitations, and Design Alternatives*, Information Resources Press, Arlington, Va 1983, 189pp.

C.L. Borgman, D. Moghdam and P.K. Corbett, *Effective Online Searching, A Basic Text*, Marcel Dekker, New York, 1984.

C.L. Gilreath, *Computerized Literature Searching: Research Strategies and Databases*, Westview Press, Boulder London, 1984, 176pp.

T.C. Li, *An Introduction to Online Searching*, Greenwood Press, Westport, Conn. 1985, 289pp.

There are also a few journals devoted to computerized searching, e.g. *Database*, *Online*, both being published by Online Inc.; *Online Reviews* published by Learned Information Ltd.; *Information Services and Use*, published by North-Holland.

Many data bases are avilable for online searching (about 3,000 different data bases are available to the public, 15–20% of them are of interest to the chemist). Most data bases are available for online searching through distributors or vendors (e.g. BRS, Dialog, Orbit in the USA; Pergamon Infoline in the UK; Telesystemes-Questel in France; CISTI in Canada; STN International in Germany, Japan and the USA). However, there are a few data bases or data base search capabilities which are only available to the public directly from the producer. These are data bases or parts of data bases which are kept by the producer himself and are not licensed to vendors.

Some of the data bases are available exclusively from one distributor [e.g. *CASSI* which is the online version of *Chemical Abstracts Service Source Index* is available only via Orbit; *CAOLD* which is the online version of pre-1967 *CA* is available only via STN; *IRIS* (Infrared Information System) which is produced by the Sadtler Research Laboratories, is available through University Computing Company]. Other data bases are available through a number of vendors (e.g. the bibliographic files of the *CA* are available from nine different vendors). Sometimes there are variations in the same data base availability among different vendors. Such is the case with the *CA* bibliographic files. The coverage of BRS, CISTI and JICST is only from 1977, while the coverage of the other six vendors (Data-Star, Dialog, ESA-IRS, Orbit, STN International and Telesystemes Questel) goes back to 1967. The updating of the *CA* bibliographic files varies among the vendors. Whereas BRS and CISTI update their files once a month, all other vendors update this particular file every two weeks. The Chemical Abstracts Service (CAS) does not license the abstracts to any vendor. However, they could be searched and retrieved from STN International which is partly owned by the CAS.

Detailed information about available data bases, vendors, producers, etc., could be found in the latest editions of *Computer-Readable Databases: A Directory and Data Sourcebook* published by the American Library Association and Elsevier Science Publishers or *Directory of Online Database* published by the Cuadra Associates Inc. and Elsevier Science Publishers.

In order to obtain the best results in a manual search one needs to have a thorough knowledge of the data base being searched, e.g. coverage, indexing policy, limitation, special features. However, for an effective online search one needs, in addition, to have a good knowledge of the online system being used. One should be familiar with the software used by the distributor; this can usually be found in the various manuals produced and distributed by the different online systems and/or the data base producers. For example, the CAS published three versions of its *CA Search for Beginners*—BRS, Dialog and Orbit version which is an introduction for online searching of the *CA* bibliographic files. It also published *How to Use CA Search: An Instructional Manual on On-Line Access to CA Search*, which is a more advanced manual which compares search strategy and search language of the above three systems.) The various suppliers and distributors conduct training sessions in various locations, and one should try and take part in these sessions. One should remember that the online suppliers continuously make changes and improvements to their systems. The user should keep track of these changes via the newsletters which are issued by the various suppliers.

The pricing policies of online searching, although based on the principle of 'pay as you use', are not so simple. There are several components to the prices and they are combined in various ways. Furthermore, prices are subject to change at short notice.

The basic component of the search cost is the 'connect time', which is charged according to the time the user is connected to the system computer (the host computer). This price varies from $35 to $300 per hour depending on the data base and the system used. To this price one has to add communication charges, which vary from a few cents to $30 per hour. The charge depends upon one's geographic location in relation to the host computer [one has to pay for the traffic volume (amount of data transferred) when data communication systems are used internationally] and the acession route used (e.g. local direct dialing, long distance direct dialing, telex, data communication system—Isranet, Tymnet, Telenet).

There are various variations in the basic connect time. Most of the vendors give reduction to large-volume users. Some data bases have two different connect time charges, a lower one to users who subscribe to their paper product and a higher one to non-subscribers (e.g. *Scisearch* which is the computerized version of the Science Citation Index, *WPA* which is the computerized version of the Derwent Patent Index). There are cases in which an annual subscription fee has to be paid in order to gain access to the data base (e.g., the Chemical Information System, CIS). In other cases one has to subscribe to a minimum usage within a specified period (e.g. *IRIS*).

Additional charges are added if the user decides to store selected data from the data base together with his own data in a 'private file'. Search strategies can also be stored for a minimal fee on the host computer.

Most data bases charging for online types ($0.02–$2.50 for type). Whenever there are more than a few citations retrieved, one should consider offline prints of the citations. Their cost is somewhat higher than the online types ($0.06–$10.00); however, there is no connect time and communication charges. The online types and offline print charges vary according to the data base used and the amount of information which is typed or printed.

One has to keep in mind that there is also a capital cost of the computer terminal or the microcomputer which is used for the communication with the host computer. In addition one needs equipment to couple the terminal (or the micro) to the communication network. This equipment (a modem or an acoustic coupler) may be built into the terminal (or the micro) or may be purchased separately. The cost of the hardware varies widely depending on the manufacture and/or the special features added. It can be as low as $1000 for a simple character terminal with an acoustic coupler, or as high as $8000–$10,000 for a micro-computer with an intelligent graphic capability, large built-in memory (640 KB), high-performance data communication system, a Winchester disc, and a high-speed printer.

An online (or computerized) search is very similar to a manual search using printed indexes. One should prepare one's search strategy in advance in a written form, specifying clearly the search objectives and search terms using the three basic logic Boolean operators—AND, OR, NOT. Search aids which are available in a printed form and/or microform should be used as much as possible in the preparation of the search strategy (when preparing an online search in the *CA* bibliographic files one should use one or more of the following *CA* search aids: *Natural Language Term List, Rotated Title Phrase List, CA General Subject Index Heading List, 11th* and/or *10th* and/or *9th* and/or *8th Collective Index Substructure Searching via Nomenclature Manual and Search Aids*). There are also various online searching aids which should be used (some of them are not available in a printed form).

After running the search, the user should scan online a small sample of the data retrieved (usually two or three citations are selected at random). According to this sample, one should decide whether any changes to the search strategy are necessary. Once the user is happy with the sample results, all of the data retrieved can be typed online or printed offline.

Online searching is usually carried out in one's own institution (usually in the library) by the chemist himself, by the librarian, or by an information specialist. Experience has shown that the best results of chemical online searching are obtained when the search is carried out by a chemical information specialist (or at least an information specialist with a good chemistry background) in cooperation with the end user (the chemist). There are several reasons for this: the end user and the specialist have a common language; the chemical information specialist is familiar with the printed forms of the various chemical data bases (especially the *CA*), and is also familiar with the various online searching systems used (e.g. BRS, Dialog, STN International). Thus he could decide which are the best data bases to search and in which system the search should be run.

In order to have a thorough knowledge of the online system, one must study the various manuals which are produced for the system. More competence is gained by attending training, refreshers, and updating seminars given by vendors and/or producers. However, one obtains expertise in using the system simply by using it.

One can conduct an online search even when there is no online search system at one's disposal. In such a case one has to rely on the services of a public library, an information consultant, or an information centre.

The great disadvantage of using a public library service is that usually the system operator does not have any chemical knowledge. However, there are various chemistry libraries (e.g. The Royal Society of Chemistry Library) which provide the chemical community with an online search service, conducted by a chemical information specialist.

The information consultants are private, independent enterprises which can conduct or arrange to have conducted computerized and/or manual searches.

The information centres provide a wide variety of search services. Some of their services deal with developing and maintaining one's personal current awareness program, others conduct specific search services. They are licensed by various secondary information producers (e.g. BIOSIS, CAS, ISI) to supply their customers with data derived from the producer's magnetic tapes. The various centres also provide online and offline searches using the various vendors' systems. Some of the other activities and advantages of using an information centre are referral and/or making contact with experts in other organizations, obtaining difficult to locate documents, and evaluating the retrieved data.

One should be aware of the existence of nearby information centres and use their services whenever there is a need.

Let us go back to our search problem—what did Felix Bergmann publish in 1978 on enkephalin? The manual search was described earlier, the online search using the *CA* file of STN International is shown below:

34

⇐ FILE CA ←————————————————————————————(1)

⇐ E BERGMANN, F/AU ←————————————————————————(2)
E1   15 BERGMANN, ERNST DAVID/AU
E2   1 BERGMANN, EVA/AU
E3   34 BERGMANN, F/AU
E4   7 BERGMANN, FEDERICO A J/AU
E5   82 BERGMANN, FELIX/AU
E6   1 BERGMANN, FELIX G/AU
E7   2 BERGMANN, FRANZ/AU
E8   2 BERGMANN, FRANZ JOSEF/AU
E9   1 BERGMANN, FREDERICO W/AU
E10   8 BERGMANN, FRIDA/AU
E11   1 BERGMANN, FRIEDRICH WILHELM/AU
E12   6 BERGMANN, FRITZ/AU

⇐ S E3 OR E5 ←————————————————————————————(3)
            34 "BERGMANN, F"/AU
            82 "BERGMANN, FELIX"/AU
L1   116 "BERGMANN, F"/AU OR "BERGMANN, FELIX/AU

⇐ S ENKEPHALIN ←——————————————————————————(4)
L2   4024 ENKEPHALIN

⇐ S L1 AND L2 ←————————————————————————————(5)
L3   4 L1 AND L2

⇐ S L3 AND 1978/PY ←————————————————————————(6)
        421732 1978/PY
L4   2 L3 AND 1978/PY

⇐ D L4 1-2 BIB ←————————————————————————————(7)
L4 ANSWER 1 OF 2 ←——————————————————————————(8)

AN CA90(23):180078w
TI The effects of narcotic agents on enkephalin levels in the rat brain
AU Weissman, Ben Avi; Bergmann, Felix; Altsteter, Raya
CS Dep. Pharmacol., Israel Inst. Biol. Res.
LO Ness-Ziona, Israel
SO Dev. Neurosci. (Amsterdam), 4(Charact. Funct. Opioids), 297-8
SC 1-5 (Pharmacodynamics)

DT J
CO DNEUDS
PY 1978
LA Eng.

L4 ANSWER 2 OF 2 ←————————————————————————————(9)

AN CA90(19):145718h
TI In vivo interaction of morphine and endogenous opiate-like peptides
AU Bergmann, Felix; Altstetter, Rayah; Weissman, B. A.
CS Hadassah Med. Sch., Hebrew Univ.
LO Jerusalem, Israel
SO Life Sci., 23(26), 2601-8
SC 1-5 (Pharmacodynamics)
SX 13
DT J
CO LIFSAK
IS 0024-3205
PY 1978
LA Eng

⇐ LOG Y

COST IN U.S. DOLLARS	SINCE FILE	TOTAL ←—(10)
	ENTRY	SESSION
FULL ESTIMATED COST	2.38	2.38

DISCOUNT AMOUNTS (FOR QUALIFYING ACCOUNTS)

CA SUBSCRIBER	−0.32	−0.32

STN INTERNATIONAL LOGOFF AT 05:27:59 ON 07 JUL 86 ←——(11)

(1) The FILE command is used in order to select the proper CAS ONLINE file. The *CA* bibliographic data base was selected.
(2) The Expend command is used in order to display the forms in which terms are searchable. The various forms of writing F. Bergmann name were examined.
(3) The Select command is used to select the two forms of Bergmann's name (E3 Bergmann, F/au; E5 Bergmann, Felix/au) using the operator OR.
(4) The search term enkephalin was Selected.
(5) The two search terms—enkephalin (L2) and the author name (L1) were combined using the operator AND.
(6) The search results were limited to the publication year 1978.
(7) The Display command is used in order to obtain the search results as a hard copy online. The answers requested are 1 and 2, in answer set L4, in a bibliographic formate. When working with a PC the search results could be stored on a floppy disk and manipulated at a later stage. Search results could be

obtained offline by using the command PRINT.
(8) A full bibliographic record of the first article Felix Bergmann published on enkephalin in 1978.
(9) A full bibliographic record of the second article Felix Bergmann published on enkephalin in 1978.
(10) COST information is displayed automatically when you logoff (LOG Y) or change files. The search cost came to $2.06 to *CA* subscribers and $2.38 to nonsubscribers.
(11) Search time can be calculated from logon time and logoff time (or starting and ending time in a specific file). The search time was 1 min. and 32 sec.

3.4. Locating material outside the library

There are very few libraries which hold most (over 90%) of the scientific books and journals which have been published during the years. Furthermore, from time to time there is a need for some government agency report or document, for a patent, or for some archival material, which are not held by the average library. How does one locate such material which is not held by an institute library and/or by an accessible library?

3.4.1. Locating a book

The first step in locating a book which is unavailable in one's library is to inquire at the various other libraries in the immediate vicinity. If this does not help one should consult the Union Catalogue of the nearest bibliographic centre. As a last resort, one should use the National Union Catalogue. In case the desired book is unavailable in one's own country, one may use the services of other National Libraries (e.g. the British Library Document Supply Centre—BLDSC—formerly known as the British Library, Lending Division—BLLD—may supply photocopies from its own stacks as well as from other major libraries with which it has special arrangements). In case of rare out-of-print books, the best place to check would be a library covering the appropriate subject matter.

3.4.2. Locating a journal article

One should first check the various Union Lists (lists of the various journals with their location), first the local (whenever there is one), then the national list. There are cases in which the desired journal may not be available in one's own country. Whenever this happens one should turn to the *CASSI*, which lists libraries around the world (326 libraries in the USA and 72 libraries in 27 other countries) that hold the various publications cited in the *CA* since its inception in 1907. It also lists journals covered by Biosciences Information Service (BIOSIS), Engineering Index (EI) Inc., and the ISI, as well as all titles covered by *Chemisches Zentralblatt* from 1830 to 1940 and by *Beilstein's Handbuch der Organischen Chemie* up to 1965. Quarterly supplements bring the work up to date (each fourth supplement being a cumulative annual one), followed by a cumulative new edition every five years.

Once a holding library has been located a photocopy of the desired article could be ordered from the holding library. It is desirable to add to the request the reference source (preferably the *CA* reference) so that in case of an error, the chance of correction is enhanced. In many cases it may be worth while to use commercial services for locating and obtaining the desired article(s). Three kinds of commercial services are available.

(a) The data base producers, e.g. the CAS Document Delivery Services (CASDDS), provides documents cited in the CAS publications and services for the past ten years (Soviet documents for the last fifteen years, most foreign chemical patents from 1977 to the present, US patents from 1971 to the present); ISI's The Genuine Article (formerly OATS) provides Original Article Tear Sheets or photocopies of all journals covered by the ISI data bases since 1978. (The latest edition of the *CASSI* indicated which of the journals are covered by the Genuine Article service of the ISI and/or by the University Microfilms International Article Clearing House (UMIACH) whose coverage extended from 1978 to the present.)

(b) Libraries and Information Centers—e.g. BLDSC; Colorado Technical Reference Center (CTRC); Delft University of Technology Library (DELFTLIB); Technical Research Centre of Finland (VTTINF).

(c) Private organizations—e.g. Information On Demand (INFO); Information/Documentation (INFODOC).

Let us see how one can locate a desired article that was published in a journal which is not available in one's own library. Let us look at an Israeli chemist dealing with the recombination of synthetic and semisynthetic analogues of the A and B chains of insulin who is interested in the modified procedure of Yung-Chin Chiang for the recombination of natural A and B chains. This procedure was published in *Sheng Wu Hua Hsueh Yu Sheng Wu Wu Li Hsueh Pau, Peiping*, **3**, 176–180 (1963) (*CA* 59 : 11828e). Checking the Union List of Serials in Israeli Libraries the chemist found (as he had suspected) that this journal is unavailable anywhere in Israel. Using the *CASSI* he found that the desired volume of the journal is held by nine libraries in the USA (e.g. University of California, Berkeley; MIT; Wayne State University; New York Public Library; Brown University), two in Canada (e.g. Canada Institute for Scientific and Technical Information—CISTI), and one each in France and Russia. Not only did he obtain the addresses of all the 13 libraries from the *CASSI*, but he also found that the papers published in the desired journal are written in Chinese with Chinese and English abstracts. All that was left to our chemist to do was order a copy of the desired article from one of the libraries he located or apply to a commercial information organization with good connections to one of the above libraries (e.g. FIND/SVP Document Retrieval Services has its offices next door to the New York Public Library; Information On Demand claims to have staffed source service at UC Berkeley Library; Information Store lists UC Berkeley Library as one of its sources).

3.4.3. Locating a patent

Although there are a few libraries that maintain files of local and/or foreign patent specifications, most libraries do not have a patent collection, and usually when in need of a patent document one has to order it. *CASSI* has a special section entitled Patent Literature, which gives considerable information about patents of countries covered by the CAS. The information includes the addresses of Patent Offices in the various countries, the fees they charge for a copy of a patent, and a list of libraries (out of the 398 covered by the *CASSI*) that hold patent collections of the various countries.

Although copies of patents can be obtained from the Patent Office for a nominal fee, or from the various libraries that hold them, none of these institutions can match the service and speed of commercial organizations. CAS-DDS supply patent documents; however, their service is limited to patents appearing in the *CA* and does not go back before 1977 (for non-US patents) or 1971 (for US patents). BLDSC supply patent copies from its collections (there are about 24 million patents from all countries in the collection). Some of the best known commercial patent organizations are Derwent Publications Ltd., IFI/Plenum Data Company, and International Patent Documentation Center (INPADOC).

3.4.4. Locating a thesis

Theses can usually be borrowed or obtained as photocopies, microfilm, or microfiche from the library of the institution where the work was carried out. This is simple once all, or most, of the bibliographic data for the desired thesis are available (author's name, thesis title, institution, year). How can one trace a thesis when only part of the data is available or when one is looking for a thesis on a certain subject without any bibliographic information at all?

Master theses submitted to American universities can be traced by using either *Master Theses in Pure and Applied Science*, or *Master Abstracts*. *Master Theses in Pure and Applied Science* is published annually by the Thermodynamic Properties Research Center of Purdue University. *Master Abstracts* is published quarterly by University Microfilm International (UMI) and includes at present bibliographic data and abstracts from a little over 100 universities, most of them in the USA. A cumulative Author and Subject Index is included in the December issue of each year.

Information about American dissertations, and to some extent about dissertations from various other countries, can be found in *Dissertation Abstract International B, Science and Engineering*, which is published monthly by UMI. *Dissertation Abstracts International* covers Doctoral theses submitted to about 360 universities (over 90% of the institutions are in the USA and Canada), each issue having only an Author Index while the last issue of each volume contains a cumulative Author and Subject Index for the whole volume. Tracing American dissertations back to 1861 and dissertations reported in *Dissertation Abstracts International* can be done by using *Comprehensive Dissertation Index*, which originally covered dissertations up to 1972 but is kept up to date by supplements. Volumes 1–4 of *Comprehensive Dissertation Index* are devoted to theses in chemistry. *Dissertation Abstracts* can be searched online; it is interesting to note that while the computerized files are held by two vendors—one (BRS) holds only the *Dissertation Abstracts* starting from 1861, while the other (DIALOG) holds the *Dissertation Abstracts* from its begining (1861) as well as the *Master Abstracts* from 1961 in the same file.

Similar sources for tracing theses are available in other countries, e.g. *Index of Theses Accepted for Higher Degrees in the Universities of Great Britain and Ireland, Catalogue des Theses Escrits Academique, Jahresverzeichns der Deutschen Hochschulschriften*.

Copies of dissertations reported in *Master Abstracts* and in *Dissertation Abstracts International* can be obtained rapidly for a reasonable fee as a photocopy or in microform from UMIACH.

3.4.5. Locating technical and/or national/international reports

Documents of national and international agencies can be obtained directly from the agency concerned. Unclassified US Government-sponsored research reports can be obtained from the National Technical Information Service (NTIS) for a modest fee. Bibliographic data as well as ordering information concerning the above documents can be found in the *Government Report Announcements and Index* which is published by the NTIS. Various reports reported in the *CA* are followed by information which enables the user to order a copy of the report (Fig. 3.5).

and vehicle to solid products having high aromaticities occurred due
to the lack of H in the reaction system.
103: 163178y Combustion tests – extended peat sods produced
from peat bogs of Lac Saint–Jean region, Quebec. Marsan, A.
(Natl. Res. Counc. Canada, Halifax, NS Can.). *Report* 1983,
NRCC-23015, 25 pp. (Eng). Avail. NRCC. From *Energy Res.
Abstr.* 1985, 10(12), Abstr. No. 22358. In combustion of low moisture
peat, moisture content (m.c.) (15%). steam prodn is ~7 lbs per lb

Figure 3.5 A *CA* entry to a report

40

The availability code identified the source from which the document can be obtained. Explanations of availability codes are given in the introduction of the latest first *CA* issue volume. NRCC stands for the National Research Council of Canada, Ottawa, Ontario, K1A 0R6, Canada. The accession number of the document is also given (NRCC-23015). This number should be used when ordering the report.

3.4.6. Locating deposit material

Material deposited in deposit centres are reported in the deposit centre bulletin or journal (e.g. material deposited in the deposit centres of VINITI are reported in its *Catalogue of Deposited Manuscripts*). They are abstracted and indexed by the various secondary information services. Here again, the bibliographic data in those abstract journals is followed by ordering information (Fig. 3.6.).

translated.
103: 24570g New data on types of structures and formations in deposits of natural-gas fields in the external zone of the Ciscarpathian Foredeep. Shcherba, V. M.; Shcherba, A. S. (USSR). *Deposited Doc.* 1984, VINITI 4926-84, 22–46 (Russ). Avail. VINITI. Geol. features are described of the Miocene sediments (sandstones and claystones) of Ciscarpathia. These

Figure 3.6　A *CA* entry to a deposit document

The document could be obtained from VINITI—All-Union Institute of Scientific and Technical Information, TsIONT, Oktyabr'ski'pr 403, 140010 Lyubertsy 10, USSR. (The accession number VINITI 4926-84, 22-46 should be used when ordering the document.)

3.4.7. Locating proceedings of scientific meeting

Locating proceedings of scientific conferences is easy if they have been abstracted and indexed by the CAS. The first step is to find out whether this has been done, by using the *CASSI*. If the result is positive, *CASSI* gives full bibliographic information of the proceedings volume as well as information about the libraries which hold the desired volume.

If the proceedings have not been processed by the CAS the search is more complicated. The most comprehensive list of scientific conference publications is the BLDSC monthly *Index of Conference Proceedings Received*. Each issue contains a list of conference proceedings with their bibliographic information and a Keyword Subject Index. There are annual cumulations; a retrospective search could be carried out in a much shorter time using the *Cumulative Index of Conference Proceedings Received 1964–1981* (which is available in a microfich form only) or the *Index of Conference Proceedings*

Received 1964–1973 (which is available in a printed form only). Using the above index one should remember that it is a list of proceedings received by the BLDSC; thus a big delay could occur from the publication of the proceedings till it appears in the *Index* (as a matter of fact, till its acquisition by the BLDSC) and that all proceedings appearing in the *Index* could be obtained from the BLDSC.

More bibliographical data concerning conference proceedings can be obtained by using the *Directory of Published Proceedings Series SEMT*, which covers science, engineering, medicine and technology. It appears ten times a year and the meetings are arranged chronologically by date of meeting, giving the location, the name and sponsors of the meeting, as well as full bibliographic information about the published volume. Each issue contains Subject/Sponsor, Editor, and Location Indexes and there is a cumulative annual index.

Index to Scientific and Technical Proceedings (ISTP) is a publication of the ISI which enables the user to obtain, among other things, bibliographic verification of the proceedings volume he is seeking. It is a monthly publication with a semi-annual cumulation covering the proceedings of over 3,000 of the most important scientific meetings in the various disciplines each year. It covers the proceedings whether they were published as a book, a report, a preprint (when this is the only publication product of the meeting), a journal issue, part of a journal issue, or in any other printed form. Each issue of *ISTP* contains, in addition to the Content of Proceedings, six different indexes. The Content of Proceedings gives complete bibliographic information about the proceedings and the papers they contain. Entries are arranged in ascending order by numbers, which are referred to in the various indexes. There is an indication when proceedings papers could be obtained through the ISI Genuine Article service. The various indexes are the Category Index (dividing the proceedings into various topics of interest—about 700 topics are used), Permuterm Index, Author/Editor Index, Sponsor Index, Meeting Location Index, and Corporate Index.

3.5. Translations

The dominant language in science today is English, over 60% of all primary scientific information published today being in this language. Even in countries where English is not a native language scientific publications are published in this language (e.g. Czechoslovakia— *Collection of Czechoslovak Chemical Communications*; France— *Nouveau Journal de Chimie*; Israel— *Israel Journal of Chemistry*; Japan— *Chemical Letters*). Most chemists are capable of coping with scientific literature in English, either because English is their mother tongue or because they studied English as a second language at school and/or it was a prerequisite for their admission to an undergraduate or graduate course.

What does one do when one has to read a paper written in a language other than English that one does not understand? The simplest solution is to have a translation made, but this is very expensive. First, therefore, one should try to locate a suitable English abstract or summary. This may be found in the original journal (many non-English and all multi-language journals publish abstracts or summaries in several languages, of which one is usually English, e.g. *Chemische Berichte, European Polymer Journal*) or in an abstract journal (e.g. *CA*). The abstracts give some indication of the value and quality of the material, and whether it is worth while pursuing further. Once the abstract or the summary has indicated that the paper is of interest, one should try to obtain the material in a language one can understand.

One should find out if there is a translation of the publication in question. A large number of journals (mainly Russians) are translated cover-to-cover into English. Using *CASSI* one can find if a cover-to-cover translation of the journal concerned exists, in which case there is also a cross-reference to the translated journal. A list of journals being translated cover-to-cover has been published by the BLDSC and the International Translation Center—*Journals in Translation*, 3rd ed., 1982. If there is no cover-to-cover translation of the journal in existence, the next step is to find whether the specific paper itself has already been translated. This could be done using information obtained from a translation centre.

In most developed countries there are translation centres whose function is to collect translations into their own language, e.g. the National Translation Center at Chicago's John Crerar Library in the USA; the Centre National de la Recherche Scientifique in France; the International Translation Centre (ITC), which was founded in 1960 under the auspices of the Organization of Economic Cooperation and Development and was originally called the European Translation Centre. The International Translation Centre collects translations from the 'difficult' languages (Eastern European, Far and Middle Eastern Languages) into any of the Western European languages. It is located at Delft, The Netherlands (101 Doelenstraat 2611), and at present it has 12 member countries (Belgium, Denmark, France, West Germany, Greece, The Netherlands, Norway, Portugal, Spain, Sweden, Switzerland and the UK).

All the translation centres publicize their collections in special publications, where one can find whether the paper one is interested in has already been translated. When using these publications, one should keep in mind that there is no relationship between the date of publication of the original paper and the date of appearance of the translation. One should therefore check the translation centre publications from present back to the original publication date. The two most important publications of this kind are: *Translations Register-Index* and *The World Index of Scientific Translations*.

The *Translation Register-Index* has been published since 1967 by The National Translation Centre. It appears twice a month and cumulated quarterly and annually. The *Register* part contains information about new accessions by the Centre—arranged in subject order. The *Index* part is arranged in alphabetical order of the journals from which the translations are made (citation index) and contains translations reported in the *Register* section as well as information about translations collected by the BLDSC and US Government-sponsored translations which are reported originally in *Government Report Announcements*. The period prior to 1967 is covered by *Consolidated Index of Translations into English* which is organized on similar lines and lists over 160,000 translations.

The *World Index of Scientific Translation* is published by the International Translation Centre and is issued monthly with quarterly, yearly and five yearly cumulations. It is arranged on similar lines to the *Translation Register-Index*; the first part of the publication lists translation accessions according to subjects and the second part is arranged according to a citation index.

All of the above 'translation clearing houses' deal with translations in all fields, although science and technology are the dominant ones. There are a few organizations which prepare, collect and announce translations in specific fields (e.g. Euroatom in the nuclear energy field), but these translations become available sooner or later through the National Translation Centre.

If one cannot find a ready-made translation, one should try to locate an 'equivalent' to the desired material. For a foreign-language paper, some or most of the data required might be found in another paper written in English by the same author (it should not be hard to identify such duplication). Another possibility is that a review paper or book covering the subject also describes adequately the specific work one is after. For a foreign patent, one should try to find any corresponding identical patents granted in English-speaking countries by using the *CA* Patent Index (or Patent Concordance for patents granted before 1977). This procedure should be adopted first when one is looking for a translation of a patent.

If no translation or an equivalent can be located one should attempt some translation work with the help of a dictionary. Some recommended dictionaries are:

A.M. Patterson, *A French-English Dictionary for Chemists*, 2nd ed., Wiley, New York, 1954, 476 pp.

A.M. Patterson, *A German-English Dictionary for Chemists*, 3rd ed., Wiley, New York, 1950, 541 pp.

These two dictionaries are very extensive in their coverage and have been used as standard references for many years. Chemistry and related areas such as biochemistry, biophysics and physiology are covered in the second revised edition of L. DeVries and H. Klog's *Dictionary of Chemistry and Chemical Engineering* (*Worterbuch der Chemie und der Chemischen Verfahrenstechnik*), Verlag Chemie, Weinheim Bergstr., 1978–79, 2 Vols. (Vol. 1, *German-English*, 851 pp.; Vol. 2, *English-German*, 784 pp.). Recommended Russian-English dictionaries are:

L.I. Callaham, *Russian-English Chemical and Polytechnical Dictionary*, 3rd ed., Wiley, New York, 1975, 852 pp.

M. Hoseh, *Russian-English Dictionary of Chemical Technology*, Van Nostrand Reinhold, New York, 1964, 540

Pergamon Press published at the beginning of the 1970s some four-language (English, German, French, Russian) dictionaries devoted to specific topics. Among them some which are directed to the chemists:

H. Angele, *Four Language Technical Dictionary of Chromatography: English, German, French, Russian*, Pergamon, London, 1971.

H. Moritz and T. Torok, *Technical Dictionary of Spectroscopy and Spectral Analysis: English, German, French, Russian*, Pergamon, London, 1971.

The step before the last one is to try and find a colleague who can provide an oral translation on the spot. When one cannot achieve reasonable results by using any of the above methods, then the last resort is to order a translation either from a commercial translation service or from a freelance translator.

Let us now return to the example of the Israeli chemist and the 1963 Chinese paper about the recombination of the A and B chains of insulin (see p. 37). How could he obtain a translation of the paper? *CASSI* indicates that there is no cover-to-cover translation of the journal in which the paper was published into English; papers are written in Chinese with Chinese and English abstracts. Before going further, our chemist wanted to obtain an English abstract of the paper in order to decide about its importance. This could be done either by ordering a photocopy of the paper from one of the holding libraries, or from a commercial service, or simply by locating the paper's abstract in the *CA*. It would take a few weeks in Israel (or about a week in the USA) to obtain a photocopy of the original paper, but only a few minutes to locate the abstract in the *CA*. In the 7th Collective Author Index of the *CA* which covers the years 1962–67 one can find:

Chiang Yung-Chin, Tu, Y.T., and Chou, C.L.
  Increase in the activity regenerated in the resynthesis of insulin from its glycyl and phenylalanine chains, 59: 3024b
  Resynthesis of insulin from its glycine and phenylalanine chains.
  III. Increase in the level of regeneration activity, 59: 11828e.

Examining the two abstracts our chemist found that not only is the first abstract not an abstract of the paper in which he is interested, but it is an abstract of a paper written by some other workers, none of them connected with the original paper. His conclusion was that some kind of an error had occurred during the preparation of the Author Index. (Mistakes and errors, although rare, will always be found in the abstracts themselves or in the indexes. It is impossible to achieve absolute perfection in such a huge work even with all the efforts, hard work and goodwill of a devoted professional staff.) The second abstract was the one of the desired paper. He learned from it that the increased regeneration activity was caused by some experimental modifications: avoidance of urea, exact control of temperature and time of the reduction and the reoxidation, and quantitative recovery of the reduced chains by their participation at pH 3.8. Furthermore, at the beginning of the abstract there was a reference to *Scientia Sinica (Peking)*, **12**, 452 (1963). The original paper seemed likely to be useful and he decided to check whether an existing translation was available. The *Translation Register-Index* and *World Index of Translations* (including the *Consolidated Index of Translations*) gave negative results, i.e. the particular paper had not been translated. He then started to look for an 'equivalent' to the paper. An equivalent would deal with the same subject—recombination of the A and B chains of insulin or maybe just insulin recombination. Using the 7th Collective Subject Index of the *CA* he found three entries under insulin recombination: 56:1521g; 59:3024a; and 59:11828e. The last entry is the abstract of the original work; as the paper he was looking for was published in 1963 he ignored the first entry and checked the second. Examining the *CA* abstract 59:3024a he realized it was the article appearing at the begining of abstract 59:11828e and it looks as if indeed it was the 'equivalent' he was looking for: 'Further increase in the activity regenerated in the resynthesis of insulin from its glycyl and phenylalanyl chains' Rong-Qing Jian, Yu-Cag Du, and Chen-Lu Tsou (Acad. Sinica, Shanghai, China) *Sci. Sin. (Peking)*, **12**, 452 (1963) (in English). The abstract started: 'Improvements in exptl. conditions resulted in the regenerated activity [*ibid*. **10**, 84, 332 (1961)]. Temp. during reoxidation was important. Splitting of the disulphide was correlated with a drop in activity...'.

It seems to the chemist that the abstract of the original paper together with the *Sci. Sin. (Peking)* papers [he knew from abstract 59:3024a that the *Sci. Sin. (Peking)* is written in English] would give him all the information he was after. Although neither his institution's library nor any other Israeli library holds this Chinese journal, he should be able to locate a library that does hold it using the *CASSI* and order a photocopy of the desired articles.

CHAPTER 4

Keeping Updated—Current Awareness Programme

Today, in the midst of an explosion of scientific knowledge, how can a chemist keep track of current trends and developments in science in general and in chemistry in particular? How can he keep himself up-to-date in his own specialized area of research? How can he locate the new information he is interested in from well over half a million items (e.g. journal articles, reports, papers presented at scientific meetings, theses, books, patents) annually? In order to overcome this problem, one has to develop one's own current awareness programme, based on a small number of core journals and on the products of the various abstracts and indexing services (e.g. CAS, ISI).

4.1. Preparing a current awareness programme

One should usually spend 5–10 hours a week keeping up-to-date, time which is spent either in the library or at home. When preparing such a programme, one should also think about its cost. Some of the tools used in a current awareness programme, particularly those based on computer search, cost money, and personal copies of journals, magazines, and books have to be paid for. About $750–1200 a year should cover a complete computer-based awareness programme for a research chemist.

The first thing each chemist preparing his current awareness programme should do is to define clearly his current professional interests—subjects about which he wishes to keep himself informed, people and organizations whose activities he would like to follow, etc. Those definitions (or statements), which are also known as a profile, are a very important tool in the current awareness programme, especially in the parts which are based on computer usage.

There are three levels in one's current awareness programme. The first is the general level, dealing with new trends in science in general and in chemistry in particular. The second is the general chemical level, dealing with research results in broad areas of chemistry (e.g. analytical chemistry, biochemistry, organic chemistry, physical chemistry). The third is the most important level in the whole programme, the specific level, dealing with specific subjects and topics about which one wishes to keep up-to-date. However, one has to keep in mind that there are no sharp borders between these three levels.

A chemist must decide which of the scientific journals he will read or scan cover to cover, and which he will follow with the aid of information services. One should bear in mind that a list of periodicals can be changed (the same is true about a list of professional interests)—a scientist should be open to changing his reading list and his profile, to adding new journals and/or new statements to his lists, and removing journals which become of less use and/or statements which are of no interest any more owing to changes in publication policy or because of changes in one's research interests.

4.2. The general level of keeping up-to-date

On the first level, one should read one or two interdisciplinary science journals that will provide a general outlook on trends and perspectives in the natural and/or technical sciences (e.g. *Science, Nature, New Scientist*). One should also read one or two general informative chemical journals that will give an overview of the various areas of chemical research, national and international science policies, technology, education, marketing, production, etc. (e.g. *Chemical Age, Chemical Engineering, Chemical and Engineering News, Chemical Week, Chemistry and Industry, Chemical Marketing Reporter, Chemtec*).

One should keep in mind that some of the journals mentioned above also report original research results (e.g. *Science*—interdisciplinary research articles with emphasis on the life sciences, *Nature*—original reports in the life and physical sciences). Others report highlights of national and international scientific meetings (e.g. *Chemical and Engineering News*), or correlate and intergrate results on certain subjects from various sources (e.g. *Chemtec*). Most of the publications which belong to the general level are devoted to specific aspects and are oriented towards specific reader populations. This is the reason that the choice of journals at this level is determined mainly by the kind of work one does (research, design, development, science management, marketing, sales, etc.).

4.3. The semispecific level of keeping up-to-date

On the second level one should read one or two of the 'prestige' general research journals in chemistry. These are journals which publish papers that appeal to readers in more than one speciality or papers that give sufficient significant results to attract the interest of specialists in other fields (e.g. *Journal of the American Chemical Society, Journal of the Chemical Society, Chemical Communications*). One should also read the general chemical journal which is published or sponsored by one's own country's chemical society (e.g. *Chemische Berichte, Canadian Journal of Chemistry, Bulletin of the Chemical Society of Japan, Israel Journal of Chemistry*). On this level one should also read a few journals which publish experimental and/or theoretical results in all aspects of a general area in chemistry. An organic chemist should read some of the general journals in organic chemistry (e.g. *Journal of Organic Chemistry, Tetrahedron, Tetrahedron Letters, Synthesis, Justus Liebigs Annalen der Chemie, Journal of the Chemical Society— Perkin Transactions I* and *II*). A biochemist should read some of the general journals in biochemistry (e.g. *Journal of Biological Chemistry, Biochemical Journal, Journal of Biochemistry, Biochemistry, Biochemical and Biophysical Research Communications, FEBS Letters*, the biochemistry section of the *Proceedings of the National Academy of Sciences of the USA*). The same holds for a physical chemist, an inorganic chemist, an analytical chemist, etc.

In this level one should read the semispecific journals which are devoted to a specific topic within a discipline. An organic chemist working on heterocyclic compounds should read: *Heterocyclic Chemistry, Journal of Heterocyclic Chemistry*. An analytical chemist interested in chromatography should include in his journal list the *Journal of Liquid Chromatography* and/or *LC/GC: Magazine of Liquid and Gas Chromatography*.

Somewhere between the second and third level one finds the speciality journals which are journals that cover a very specific area of reasearch, usually from the interdisciplinary view point (e.g. *Journal of Fluorine Chemistry*); a chemist will read one or more of these journals, depending upon his interests. One interested in peptide synthesis will read, among other journals, *Bioorganic Chemistry* and *International Journal of Peptide and Protein Research*. A scientist dealing with the problem of the origin of life will include in his journal list one or more of the following journals: *Biosystems, Journal of Molecular Evolution*, and *Origin of Life*.

4.4. The specific level of keeping up-to-date

The third level, the specific level, is the most important in the whole current awareness programme. This level represents the areas in which one wants to keep oneself up-to-date as completely as possible. This level corresponds closely with one's special interests and areas of research. At this level one is using mainly secondary sources as well as products of the various information services.

There is a large number of information services which produce and distribute various abstracting and indexing products. One has to keep in mind that all those information services are not equal. Some index only journals, while others include journals, dissertations, patents, etc., in their coverage. Some are covering journals from cover to cover; others cover only selected papers in journals. Abstracts are provided in some but not in others. There is a difference in the way information is indexed, indexing being done in one or more of the following methods: predetermined subject categories, keywords in context (KWIC), natural language, citation index, etc. One should be familiar with the coverage and indexing policy of the information products one is using.

Bradford's law states that most of the important material published, is published in a relatively small number of journals. Indeed, we have seen before that 25% of all published articles in chemistry appear in a total of 90 journals, 50% in 450 journals, and 75% in about 1600 journals (0.45–0.50% of the original chemical research results are published in one journal—*Journal of the American Chemical Society*). How is one able to locate articles of importance in the current literature? The simple answer is to scan through the 100–200 core journals in one's field. However, this is really impractical unless one decides to spend a large percentage of one's time in the library, and the library decides to have a collection of well over 5000 current journals.

4.4.1. Scanning journal titles

Two secondary journals help the chemist to skim selectively through the core literature: *Current Contents* published by the ISI, and *Chemical Titles* published by CAS.

Current Contents is a weekly publication published in seven different editions, five covering various disciplines of the life and physical sciences, one covering the social sciences, and one covering the arts and the humanities. A chemist should use one of the following editions, depending upon his interests: *Current Contents, Physical Chemical and Earth Sciences; Current Contents, Engineering Technology and Applied Science;* or *Current Contents, Life Sciences. Current Contents* reproduces the lists of contents of the most important research journals in particular disciplines (it covers 700–1300 journals in each edition), followed by a Word Index and an Author Index and Directory. The publisher receives the various lists of contents immediately upon publication of the various journals (in some cases even prior to their publication), so that the time lag between the original publication of the various journals and the appearance of its contents list in *Current Contents* is never more than 2–4 weeks, usually less. In 1985 the ISI started the publication of a personal edition of *Current Contents—Custom Contents*. It is a weekly listing of contents of any journals out of the 7200 journals covered by the ISI. If you are interested in a diverse collection of journals scattered in more than one edition of *Current Contents* or in small number of extra journals not covered by the *Current Contents* edition you are using, you should order your costume copy (its basic price is $100 per year, this price covers up to ten journals, each additional journal costs $10, the basic price of any *Current Contents* edition is $257 a year).

An online version of the various *Current Contents* covering the experimental sciences (*Agriculture Biology and Environmental Sciences, Clinical Medicine, Engineering Technology and Applied Sciences, Life Sciences, Phys-*

ical Chemical and Earth Sciences) is available via BRS. The online file known as CCON consists of a three month rolling file which is updated weekly.

Chemical Titles is a chemically oriented biweekly publication of authors and keyword indexes of selected papers from about 700 journals in chemistry and chemical engineering. The publication is divided into three parts. The first part is the KWIC Index, which is a rotated title index arranged alphabetically with an entry for each keyword in the title. The index is formatted with the keywords aligned vertically with the other words in the titles before and after them. The second part is the Bibliography Index, in which the selected titles are listed in tables of contents form, arranged according to the journals covered in the issue. The journal citation precedes each 'table of contents'. The third part is the Author Index, in which all authors are listed alphabetically. The KWIC and the Author Indexes refer the users to the bibliographic citation by giving the journal CODEN and the volume and/or issue number together with the first page number of the article in question.

The CODEN system is a machine-readable method of specifying the titles of periodicals by using six characters. A CODEN is made by taking four mnemonic letters from the journal's name (e.g. *JACS — Journal of the American Chemical Society*), followed by a fifth letter. This fifth letter increases the total availability of CODEN combinations and permits the differentiation of journals with the same mnemonic letters. The sixth character is a computer checking letter (thus the CODEN of the *Journal of the American Chemical Society* is JACSAT). The system was devised and adopted by the American Society for Testing and Materials (ASTM) in 1963. Detailed information about the CODEN system and a listing of approximately 150,000 CODENs can be found in the latest edition of the *International CODEN Directory* which was published at the beginning of 1985 by CAS. It contains an alphabetical listing by titles of journals to which CODENs have been assigned together with the CODENs themselves, and an alphabetical list of CODENs together with the journals to which they were assigned. At the beginning of each year a supplement updating the CODEN data is published; each supplement cumulates the information from the previous supplements as well as giving new data. At the end of a five-year period a completely new *International CODEN Directory* is published. The Directory is available only in microfiche.

Each of the above two publications (*Current Contents* and *Chemical Titles*) has its pros and cons. Both journals serve only as a reference pointer and do not give the user immediate access to an abstract or to any information concerning the papers reported. If one is interested in one of the titles reported in those publications, one has to find the original article, order a reprint from the author (if he is using the *Current Contents*, he could locate there the author's address in the Author Index and Directory), or use the Genuine Article service of the ISI (when using *Current Contents*) or the

CASDDS (when using *Chemical Titles*). *Current Contents* is interdisciplinary in character while *Chemical Titles* is completely chemically oriented. Some of the *Current Contents* editions have quarterly indexes (e.g. *Quarterly Indexes to Current Contents Life Sciences*), thus making the *Current Contents* not only a current awareness tool but also a usable instrument for a retrospective search. When using *Current Contents* one usually scans through all the tables of contents of the various journals in the issue; using *Chemical Titles* one scans vertically the alphabetical list of keywords, examining the full titles on either side of keywords of interest; in order to obtain the corresponding reference one has to turn to the bibliographic section. One must also remember that although the number of chemical journals covered by *Chemical Titles* (700) is greater than that covered by *Current Contents*, only selected papers from those 700 journals are reported, whereas in *Current Contents* journals are covered cover to cover.

4.4.2. Abstracts and bibliographic services

In order to obtain a comprehensive current awareness coverage in broad or specialized areas, in addition to selected journal titles one should use an abstracts service or a bibliographic service. Abstracts are summaries of scientific and/or technical publications, with bibliographic information that enables the user to locate the original work. An abstract provides sufficient information for the reader to decide whether the original publication is relevant to his interests, and whether to attempt to locate the original reference.

According to the latest edition of *Ulrich's International Periodical Directory* there are a few hundreds of abstracting and indexing publications covering the primary chemical literature. Most of these publications are devoted to specific disciplines of chemistry, some of them being so specific that their circulation does not exceed 100–150 copies.

The whole chemistry abstracts literature can be divided into the following categories:

(a) Publications covering the whole field of chemistry and related disciplines (e.g. *CA*).
(b) Publications covering an interdisciplinary area which is of interest to chemists (e.g. *Fuel and Energy Abstracts*, *Pollution Abstracts*, *Oceanic Abstracts*).
(c) Publications covering disciplines in chemistry. This category is subdivided into publications which are based on data collected by the large abstracts and indexing services (e.g. the various *CA Selects* editions covering 161 topics using the *CA* data base; the various *Ascatopics* editions covering 540 different topics, many of which are of interest to chemists, using the ISI data base) and publications produced by small services covering a defined specific discipline in chemistry (e.g. *Liquid Chromatog-*

raphy Literature Abstracts and Index, Mossbauer Spectroscopy Abstracts, Nucleic Acids Abstracts, Zinc Abstracts).

4.4.2.1. Chemical Abstracts

CA is the largest abstracts service in the world and the most comprehensive one for the chemist. It serves not only as a current awareness tool by bringing to the chemist high-quality abstracts of the current literature with complete bibliographic information, but also as a working tool for information retrieval and/or retrospective searches. *CA* is published by CAS which is a division of the American Chemical Society; it was initiated in 1907 and its headquarters have been located in Colombus, Ohio, since 1909. It is a dynamic system which has been changed during the years, expanding from 11,847 abstracts in 1907 to 500,000 abstracts in 1985 with well over 5000 different issues of its publication and all of its information available on magnetic tapes with various search services.

CA monitors over 14,000 serial publications from more than 150 countries (although most of them are chemical or chemical engineering publications, there are also astronomical, medical, biological, and physical journals), patents issued by 26 nations as well as by the European Patent Organization (EPO) (under the European Patent Convention), and by the World Intellectual Property Organization (WIPO) (under the Patent Cooperation Treaty), as well as proceedings of scientific meetings, theses, national and international reports, deposit material, and books. Only material which reports new information relevant to chemistry or chemical engineering is abstracted and indexed.

Of the 14,000 journals and serials which are covered by the *CA* only 237 are abstracted and indexed cover to cover, i.e. every paper in every issue (e.g. *Journal of the American Chemical Society, Journal of the Chemical Society, Journal of Organic Chemistry*). Usually papers are selected from about 8500 serials in a calendar year and from most of the 14,000 within a five-year period.

CA is divided into 80 sections: 20 Biochemistry Sections, 14 Organic Chemistry Sections, 12 Macromolecular Chemistry Sections, 18 Applied Chemistry and Chemical Engineering Sections and 16 Physical and Analytical Chemistry Sections. Sections 1–34 appear one week and Sections 35–80 the following week. Detailed information on the coverage of each section can be found in the latest edition of the *Subject Coverage Manual—Subject Coverage and Arrangement of Abstracts in Sections in Chemical Abstracts.* The description of each section consists of four parts. The first part describes the types of information included in the Section. The second part describes the types of information excluded from the Section. The third part provides information about cross-reference policies for the Section and enables the searcher to use Sections covering similar subjects. The last part describes the arrangement of the abstracts in the Section, which is based on their subject contents. The internal arrangement in each section is journal articles, proceedings of scientific or technical meetings, technical reports, archive documents, theses, book announcements, and patents. At the end of the Section there are cross-references to abstracts of documents containing information that would justify their placement in more than one Section. An abstract can be printed in only one of the 80 Sections of the *CA*, and whenever there is some justification for placing it in more than one Section, it is placed in that one which reflects its main emphasis and is cross-referenced in the other relevant Section(s).

One cannot and should not scan through more than a few of the *CA* Sections (usually one or two). On scanning through a *CA* Section, one would find that a large number of the abstracts in it are of no importance or interest to him, while a good amount of work in one's own area of research is reported in some other Section(s).

Since the beginning of 1988 in-processing records of the *CA* were made available to the public by CAS. This information is available only online in the CASONLINE File *CAPreview*. The File contains 100,000–120,000 bibliographic records which will end in the *CA* File and in the *CA* issues. This way bibliographic records (without abstracts and index terms) become searchable eight to twelve weeks prior to their incorporation in the *CA*. Once the full record—with abstract and indexing terms added—is incorporated in the *CA*, the bibliographic record is deleted from *CA Preview*.

4.4.2.2. Non-chemistry abstracts and bibliographic services

Chemists whose interests are in the biological, engineering, medical, or physical aspects of a discipline may be able to obtain a better coverage by using the corresponding sections of *Biological Abstracts*, *Engineeing Index*, *Index Medicus* (or *Excerpta Medica*), or *Physics Abstracts* (or *Physics Briefs*), respectively.

Those interested in interdisciplinary subjects such as Environmental Sciences, Energy, etc., may be able to obtain better coverage using the corresponding abstracts services (e.g. *Environmental Abstracts*, *Environmental Periodical Bibliography*, *Pollution Abstracts*, *Energy Information Abstracts*, *Fuel and Energy Abstracts*).

4.4.2.3. Standard interest profile

If possible one should use one or more of the *CA Selects* topics or *Ascatopics* which are of interest instead of reading one or two *CA* Section(s). Those publications are products of computer searching of the CAS or ISI data base, respectively. The search is based on a standard interest profile (macroprofile) prepared by carefully selecting a set of index terms which reflect the specialized interest of a group of chemists. This is a Selective Dissemination of Information (SDI) aimed at a group of chemists having common interests.

The *CA Selects* gives the citation of the original work the way it was given in the *CA*, followed by the full *CA* abstract. The service eliminates the need to use the *CA* itself or any other secondary service in determining the relevance of the information cited. In contrast to the *CA* where an abstract may appear in only one of its 80 sections, in the *CA Selects* it may appear under all relevant topics. The *CA Selects* are published every two weeks and include material reported in all 80 Sections of the *CA* published within the last fortnight.

Although most of the abstracts from each of the topics come from only a few *CA* Sections (usually two to four) some of them are retrieved from many other sections [e.g. a typical issue of *CA Selects — Psychobiochemistry* contains about 210–230 abstracts, about 70% of the abstracts coming from Pharmacology (Section 1), Mammalian Biochemistry (Section 13), and Mammalian Pathological Chemistry (Section 14), 15% come from three other Sections, and the other 15% come from 16 different sections]. The low price of the *CA Selects* makes acquisition possible not only by small institutions and companies with small budgets, but also by the individual scientist for his personal use.

The whole set of topics has been growing since its inception. New topics have been added as well as a few topics which have been eliminated (there were 6 topics to start with in 1976, 22 in 1977, 110 in 1980, and 161 in 1987). The topics of *CA Selects* include Corrosion, Cosmochemistry, Forensic Chemistry, Herbicides, Insecticides, Organosilicon Chemistry, Radiation Chemistry, Solar Energy, Synthetic Fuels, Zeolites; a full list can be found in the latest edition of *CAS Information Tools*.

Recently, Bioscience Information Service (BIOSIS) and CAS started publication of a joint standard interest profile— *BIOSIS/CAS Selects*. This covers specific research topics that span the chemical and biological disciplines. Sixteen different topics are produced at present (e.g. Biochemistry of Fermented Foods, Biological Clocks, Endorphins, Interferon, Schizophrenia). Each topic contains abstracts and summaries selected from the BIOSIS and CAS data bases. It is a biweekly publication whose contents are selected by computer from the 750,000 abstracts and summaries published annually by the *BA*, *BA/RRM* and *CA*.

Very recently STN started to produce current awareness publications on topics at the frontier of scientific research, of which citation and abstracts are available online from STN files. The first publication contains information from the *CA* File and the PHYS File (the online version of *Physics Briefs*) is *Superconductors Updates*.

Ascatopics is a similar service provided by the ISI; in 1986 there were about 540 topics in the various sciences, of which about 15–18% were of interest to chemists. Some of the topics cover the same area as *CA Selects* (e.g. Adhesives, Liquid Crystals, Mass Spectrometry), but others deal with areas not covered by *CA Selects* (Dyes, Nitrogen Fixation, Peptide Synthesis). Its coverage is limited to the 5800 journals processed by the Science

Citation Index (SCI) data base. The fact that patents are not covered by this service reduces its value to industrial chemists. Contrary to the *CA Selects*, *Ascatopics* is a weekly publication which provides only citations to the articles (bibliographic data only), thus requiring the use of secondary services in determining the relevance of the information when the original source is unavailable. However, the advantages of using *Ascatopics* is the speed of publication (remember that copies of papers appearing in *CA Selects* could be obtained using the CASDDS while papers reported in *Ascatopics* could be obtained via the ISI Genuine Article service).

4.4.2.4. Specialized abstracts services

A chemist should also use one or more of the specialized abstracts publications dealing with his own area of interest. Thus a synthetic organic chemist could benefit from using one or more of the services offered to him by the ISI— *Current Abstracts of Chemistry and Index Chemicus* (*CAC & IC*), *Current Chemical Reactions* (*CCR*), *Chemical Substructure Index* (*CSI*)—or instead he may use the *Journal of Synthetic Methods* published by Derwent Publications Ltd., *Methods in Organic Synthesis* published by the Royal Society of Chemistry, or *SYNFORM* published by Verlag Chemie. Another possibility is to use the *CASREACT* which is available only online via STN.

Current Abstracts of Chemistry (*CAC*) is a weekly publication covering the synthetic organic chemistry literature in the form of abstracts of articles reporting new reactions, syntheses, and/or compounds. The abstracts are arranged by their source journals (each source journal has its ISI accession number). Each abstract has its own abstract number followed by complete bibliographical data—title, author(s), address(es), journal citation, and the original language of the article if not English. The abstract published is the one provided by the author, so one may find an abstract written in French or German although the title of the article has been translated into English. The abstract is followed by structural diagrams of all new compounds and by reaction flow diagrams. The original numbers for new compounds are used, these numbers being underlined. The abstracts contain two other features—the Analytical Wheel and the Use Profile. The Analytical Wheel describes the analytical techniques used in analysing and identifying the various compounds. The Use Profile informs the reader about applicaitons of the compound if any were reported in the original article. Each issue is accompanied by *Index Chemicus* (*IC*), which is the companion index to *CAC* and is incorporated into its weekly issues. This index is cumulated semi-annually and annually. The weekly index contains the following sections: Molecular Formula, Subject, Author, Biological Activity, Corporate, and Labelled Compounds. The cumulated indexes also contain a Journal Index and a Rotated Formula Index (Rotaform Index). In the Rotated Formula Index the molecular formula of all elements other than carbon, hydrogen, oxygen, and nitrogen are 'rotated'. Thus one can easily locate all compounds containing a specific element (e.g. boron, silicon). All new compounds reported in the *CAC* are cumulated and permuted monthly and annually in the *CSI*. The *CSI* can be used for locating specific chemical structures or fragments of new compounds.

CCR is a monthly publication giving detailed information on new and newly modified organic reactions and syntheses, taken from a little over a hundred journals of organic and pharmaceutical chemistry. In addition to giving information on new chemical reactions, it also gives bibliographic information on review articles dealing with organic reactions. Together with the bibliographic data of the article and the author abstract (see *CAC*), there is an experimental description, remarks about the advantages, and/or purpose of the new method, yields of all products obtained, hazard warnings, and a cross-reference to the corresponding abstract in *CAC*, followed by a reaction flow, when applicable, by highlights of the proven or potential application of the products. Each issue has its own set of indexes: Subject Index (a permuted index to new synthetic methods; some of its features are activities, catalysts, products, reaction name, reaction type, reagents, total syntheses), Author Index, Corporate Index, and Journal Index. These four indexes are cumulated annually.

Journal of Synthetic Methods is part of the Chemical Reactions Documentation Sevice which is run by Derwent Publications Ltd. The Journal is an extension of Theilheimer's *Synthetic Methods of Organic Chemistry* (see Section 9.3.3.1.). It is a monthly publication covering the chemical literature (including patents). The journal provides the organic chemist with information about novel chemical reactions, new synthetic methods, interesting applications and/or extension and modification of known reactions, improved synthetic methods, and applications of organometallic compounds and metal complexes to synthetic chemistry.

Information on novel reactions includes an abstract number, a Theilheimer Reaction Symbol (a formal representation of the chemical bond formed, the reaction type, and the chemical bond destroyed or the element eliminated—e.g. the oxidation of thioethers to sulphoxides by tetra-n-butyl ammonium periodate, shown in Reaction 4.1 [E. Santaniello, A. Manzocchi, and C. Farachi, *Synthesis*, 563 (1980)] would have a reaction symbol OS⇃S indicating that a sulphoxide bond is formed via an addition to sulphur and a Thematic group (this is a single-letter designation of a reaction type), followed by an abstract. Each abstract is prepared specially for the journal, giving a detailed description of the reaction, emphasizing the type of the reaction, the various starting materials, reagents used, and the products. There is a full bibliographic information on the original article followed by a reaction scheme. Each issue has a list of reviews and a comprehensive Subject Index which is cumulated annually. The June and December issues include as 'Supplementary References' information concerning modifications of, or improvements to, known reactions.

$$\text{(structure)} \longrightarrow \text{(structure)} \qquad (4.1)$$

Methods in Organic Synthesis is a monthly publication having 200–300 selected reactions per issue. It covers topics such as asymmetric synthesis, new protective groups, new reagents and synthons, biotechnical methods as well as broad outlines of total synthesis. Each issue contains five indexes which cumulate to form an Annual Index at the end of the year: Author Index, Reactants Index, Reactions Index, Reagents Index, Products Index.

SYNFORM compiles information dealing with total synthesis—starting materials, reagents, reaction conditions, yield, data on the structure, and stereochemistry of intermediates and end products. It serves as a reliable guide to modern methods of preparative organic chemistry.

CASREACT is an organic reaction data base covering reactions of organic substances (including organometalics and biomolecules) selected from papers reported in about 110 key journals. These journals represent the synthetic literature covered in the Organic Sections of the *CA* (Sections 21–34). The data base became available to the public via STN in mid 1988 containing approximately 20,000 records. There are over 200,000 single step reactions with about 100,000 more added each year. *CASREACT* coverage goes back to January 1985. However, yield for products are available (if reported by author) from mid 1986 only. All chemical substances reported in the data base can have five roles assigned: reactant, product, reagent, catalyst and solvent.

Chemists holding non-research positions in the chemical industry or in government agencies may need information on one or more of the following subjects: production trends, pricing policies, sales and marketing, construction of new plants and expansion of old ones, novel processes and new products, corporate labour and/or government activities, safety and pollution control, patents and licensing information, and personnel changes in private and/or public organizations. Neither *CA* nor any other scientific abstract or bibliographic journal (e.g. *Biological Abstracts*, *Physics Abstracts*, *Engineering Index*) would provide this kind of information. However, there are a few publications that do give the desired information. The two most important publications are *Chemical Industry Notes* (*CIN*) and *Predicasts Overview of Marketing and Technology* (*PROMT*).

CIN is published by CAS and gives the latest chemical business information. It covers about 100 business, trade, and industrial journals, government publications, magazines and newspapers (e.g. *Congressional Records*, *European Chemical News*, *Rubber World*, *Wall Street Journal*). It is a weekly publication divided into sections—Production, Pricing, Sales, Facilities, Products and Processes, Corporate Activities, Government Activities, and People. It contains summaries of articles and their bibliographic citations, together with issue indexes—a Keyword Index and a Corporate Index. There is an annual Index composed of the issue indexes and a Personal Name Index.

A similar publication is *PROMT* which is one of the publications published by Predicasts Inc. (other publications of Predicast are: *Predicast F & S Index United States*, *Predicast F & S Index Europe*, *Predicast F & S Index International*, *Predicasts Worldcasts*). Although *PROMT* is more limited in scope than *CIN*, its source coverage is much larger and its abstracts have greater depth than the *CIN* ones.

Chemical engineers and people involved in chemical and process engineering should use the *Chemical Engineering Abstracts* (*CEA*). The *CEA* is published monthly by the Royal Society of Chemistry. It is a current awareness publication containing 300–400 references and abstracts to recent literature in all aspects of chemical engineering (including areas of chemical engineering involved in mechanical, civil, electrical, and instrumental topics).

Chemists interested in bioengineering and/or biotechnology should be using one or more of the current awareness tools which are available in this area: BECAN (BioEngineering Current Awareness Notification) (*Instruments and Techniques in Cardiology, Electrodes for Medicine and Biology, Biochemechanics and Orthopaedics*) and CAS BioTech Update which are published by the CAS and based on the *CA* and *CIN* data bases (e.g. *Biosensors, Environmental Biotechnology, Genetic Engineering, Pharmaceutical Applications*). They may use one of the more general types of abstract publications such as *Bioengineering Abstracts* published by Engineering Information Inc., *Biotechnology Research Abstracts* published by Cambridge Scientific Abstracts, *Current Biotechnology Abstracts* published by the Royal Society of Chemistry, *Biotechnology Abstracts* published by Derwent, etc.

4.4.3. Citation indexes

There are cases in which a problem could be defined and/or identified in terms of a reference to a key article rather than a list of keywords; this is usually the case when one is trying to find the latest developments in a certain subject. Developments such as conformation, extension, testing, improvements, correlations, and applications are always connected to a key article, and one assumes that papers will cite the key article whenever they are dealing with such a development. Thus one can get information concerning new developments in a specific subject by using this approach. Are there any means by which one can use a reference as an indexing term in one's current awareness programme? Indeed, it can be done by using the *Science Citation Index* (*SCI*).

SCI is a bimonthly publication with annual and five-year cumulations published by the ISI. It is a multi-disciplinary publication covering about 5800 science journals as well as some non-journal material such as symposia proceedings, monographs, and multi-author books (collections of papers). Over 1500 journals which are covered by the *SCI* are of interest to chemists. *SCI* is composed of three parts: Source Index, Citation Index, and Permuterm Subject Index.

The Source Index is a full bibliographic description of all items published during the period covered (two months, one year, or five years) arranged alphabetically according to the first author's name. Co-authors are cross-referenced to the first author. The Source Index is followed by two supplement sections. One section gives bibliographic data to anonymous authors articles arranged alphabetically according to the journals' names in which they appear. The other section, the Corporate Index, connects authors with institutions where the work has been carried out, and is arranged alphabetically according to the institutions' names.

The Citation Index connects articles published during the coverage period with the article they cited. They are arranged alphabetically by the cited first author. The Citation Index is followed by the Patent Index, which connects the articles published during the coverage period with the patents which they cited. It is arranged according to the patent number (without any relation to the granting country) in ascending order.

The Permuterm Subject Index connects permuted pairs of words from titles of articles with the authors that have used them. These pairs are arranged alphabetically and are used as two-level indexing terms.

Some users of the *SCI* complain about the very small print size, but the use of a simple magnifying glass can solve the problem. The timelag to publication in *SCI* is on average much shorter than that in *CA*. In order to reduce this timelag to a minimum, the *SCI* changed from a trimonthly to a bimonthly publication in 1979.

4.4.4. Patent coverage

A chemist involved in industrial and/or technological work should use the patent literature as an important component of his current awareness programme. The importance of the patent literature as an information source on recent technological advances has increased since the time when many countries revised their patent laws. The fact that some countries known as rapidly patent publishing countries (e.g. Germany, Japan, France) publish patent applications without examination a short time after their filing makes the patent literature of those countries the most up-to-date source of information on technological progress.

The current patent literature provides information concerning not only technological progress but also the general progress of chemistry. The reason for this is that chemists employed by industrial firms often do not publish in the scientific literature, at least not until there is an adequate patent protection of their work. Often they do not publish in the open literature at all and the patent is the only form in which their work is reported. The patents also provide information about competitors' progress and achievements, giving some insight into their aims. They provide the industrial chemist with new approaches and ideas for solving his own problems. The individual research chemist may find in the current (as well as in the old) patent literature procedures which are of importance to his own research. While the patent laws prohibit the use of information in patents for commercial purposes by others without prior agreement of the patent holder, there are no limitations to using the procedures reported provided that they are not applied to any process intended for sale and/or profits. Hence one can use such procedures without contravening the law.

How can one keep aware of the current patent literature? The simplest answer is to read or scan the patent gazettes or journals published by the various patent offices [e.g. *US Patent Office Official Gazette* (USA), *Patentblatt* (Germany), *Patent Office Records* (Canada)]. This sounds simple, but it has its drawbacks: e.g. one can spend considerable time going through the most important patent journals, only to find that a key patent was issued in an unexpected country and/or an unfamiliar language. Further, most libraries do not receive the current publications of foreign patent offices. For a good coverage of new patent applications one should use the secondary literature.

4.4.4.1. *Patent coverage by CAS*

A lot of information concerning the current patent literature can be found in the CAS publications. Whether one is reading a specific Section or Sections of the 80 Sections of *CA* or one or more of the various *CA Selects*, one would be able to obtain information about chemical patents related to the area covered by this/those Section(s) and/or Selects. *CA* covers patents issued by 26 different countries and two international organizations. The greatest advantage of using *CA* to cover the current chemical patent literature is that it is probably the most widely available and usable secondary source. The emphasis is on the chemical side (especially on details given in the examples of the inventions) and not on the legal or any other aspects. Its accuracy is excellent and there is very short timelag to publication. On the other hand, the policy of the *CA* is not to publish any patents which do not contain new technical material. Thus divisions and reissue of patents as well as patent equivalents to ones reported earlier in the *CA* are referenced in the issue's Patent Index but not abstracted. Chemists interested in these types of patents should use another source for this part of their current awareness programme.

4.4.4.2. *INPADOC Patent Gazette*

The *INPADOC Patent Gazette*, published by the International Patent Documentation Centre (INPADOC), is an international equivalent to the national patent gazettes or journals. The Centre was founded by the Austrian Government in agreement with the WIPO and is located at Moellwaldplatz 4, A-1040 Vienna, Austria. INPADOC stores in a central data base all the bibliographic information on all patent documentation on a world-wide basis. The *INPADOC Patent Gazette* is a weekly publication which gives bibliographic data on all the patent documents from 49 countries added to the *Gazette* data base in the preceding week. It is a computer product and is published in a microfiche form using the computer on microfiche (COM) technique. The journal is composed of three parts: the classification part, the applicant part, and the numerical part.

The classification part lists all documents according to their International Patent Classification (IPC) number in ascending order. Patents are usually listed in more than one classification, thus giving the searcher a good probability of locating the desired information. The applicant part lists all patents recorded alphabetically according to inventors and applicants. This is an excellent tool for keeping informed about the current activities of particular applicants and/or competitors. The numerical part lists the various patents according to their granting countries and numbers.

The weak point of the *INPADOC Patent Gazette* is that only bibliographic information is reported and in order to obtain some idea of the content of a patent one has to go to the secondary or primary patent literature (whichever is more easily available). On the other hand, the *Gazette* lists all equivalents or patent family members, which enables the searcher who has located a difficult to obtain patent or a patent written in a foreign language to look for an easy way to obtain and/or read an equivalent.

4.4.4.3. Derwent patent publications

The most comprehensive coverage of the chemical patent literature is the coverage provided by the Derwent Publications Ltd., a subsidiary of the Thompson Organization.

A patent is first reported by Derwent in its *World Patent Index* (*WPI*). This is a weekly collection of indexes with yearly cumulations, arranged according to IPC, accession number, patent number and priorities claimed, giving information on all the inventions reported in the various countries covered by this service [Australia, Austria, Belgium, Brazil, Britain, Canada, China, Czechoslovakia, Denmark, Finland, France, East Germany, West Germany, Hungary, Israel, Italy, Japan (only chemical patents), Luxemburg, The Netherlands, Norway, Portugal, Roumania, South Africa, South Korea, Spain, Sweden, Switzerland, USSR, and USA] as well as those issued by the EPO and WIPO. *WPI* is published in four editions—general, mechanical, electrical, and chemical. These publications do not contain any abstracts and are used for watching for patents by patentee and for watching for the appearance or reappearance of an invention in another country.

The abstracts of the patents which are reported in *WPI* appeared a week later in *World Patent Abstracts* (*WPA*). *WPA* is published in 19 different editions: 12 of the editions are devoted each to a different country or group of countries, and 7 deal with different areas of technology. Some of the country-devoted editions cover patents on all subjects registered in a certain country, whereas others cover only chemical patents. The seven subject editions have a multi-country coverage.

The subject editions of *WPA* cover only non-chemical subjects. The abstracting follow-up of the chemical subjects is carried out by the *Chemical Patent Index* (*CPI*) (known prior to 1986 as *Central Patent Index*).

The abstracts of chemical patents appear first in the *Alerting Bulletins* (*Country Order*). This is a weekly publication of the *CPI* arranged by country; 12 different editions are published. Each issue contains its own indexes (Patentee, *CPI* Class, Accession Number, and Patent Number). A week later the same abstracts are published in the *Alerting Bulletins* (*Classified*), composed of 12 separate booklets arranged according to the Derwent Subject Classification. A week afterwards (two weeks after the original publication of the abstract), a more detailed abstract of all 'first disclosures' and 'basic' patents is published in the *Basic Abstracts Journals*. Those abstracts which are reported in one or more of the 12 separate journals frequently contain examples and/or drawings.

Those searching for patents in their current awareness programme should use one or more of the above Derwent journals based on their specific needs.

4.4.4.4. IFI/Plenum publications

Those interested only in USA patents and their equivalents in other countries could use the *IFI Assignee Index to US Patents*, which is a quarterly publication, or the customized weekly profiles which are known as *US Patent Profile*. It is a personalized profile to alert the user of patents issued to a specific company, new patents on specific subjects, or new issues pertinent to a specific patent.

4.4.4.5. Specific patent alert services

There are also specific patent alert services covering specialized areas of interest. An example of such a service is the *American Petroleum Institute* (*API*) *Abstracts/Patents*, which contains abstracts related to the petroleum refining and the petrochemical industry. It includes abstracts of all US patents and new pertinent patents granted by 23 other countries. While most of the abstracts are selected from material reported in the Derwent Patent Index a suppplementary group of related patents outside the scope of the Derwent Service is also included.

The API Central Abstracting and Indexing Service, which is located at 150 William Street, New York, NY 10038, USA, also publishes the *API*

Abstracts/Literature and the *Petroleum/Energy Business News Index*, thus giving a comprehensive alerting service to the petroleum chemist.

Petroleum Abstracts, produced by the Information Service Division of the University of Tulsa is another source of current patents, dealing with the exploration and production of oil. It produces 3000–5000 patent abstracts and 8000–10,000 literature abstracts annually.

4.5. Attending scientific meetings

A vital part of a current awareness programme is attending scientific or trade association meetings. These meetings enable the participant not only to obtain information well before its publication and to meet and listen to the leading figures in the various areas of science in general and chemistry in particular, but also to meet and get to know colleagues having similar interests, to renew old personal contacts, and to establish new ones.

Scientific meetings can be divided into two classes: the specialized small meetings, e.g. European Peptide Symposium, Symposium on Shock Tubes and Waves, Gordon Conference; and the big meetings, which try to cover many and sometimes all aspects of chemistry and/or biochemistry. Those big meetings are national or international in character, e.g. National Meeting of the American Chemical Society, IUPAC Congress, Meeting of the Federation of the European Biochemical Societies. The specialized small meetings are usually attended by no more than 100–200 participants. In such a meeting no parallel sessions are held, thus enabling the participants to take part in the whole meeting. They have a good opportunity to establish personal contacts with each other and to spend long hours in informal discussions. In the big meetings 2–50 parallel sessions may be running simultaneously. In these meetings, especially in the very big ones, the quality of the papers presented can be overwhelming while the opportunities for establishing personal contacts are reduced. However, there are advantages in attending big meetings, e.g. listening to lectures on general subjects as well as lectures and papers from different disciplines, visiting instrument exhibitions in order to learn about the latest developments in a particular area. Attending such a meeting would be useful only if there is careful planning. One should decide in advance which of the many plenary lectures, symposia, and contributed papers one will attend.

One may spend the whole year attending scientific meetings. In fact, attending meetings is like reading scientific journals, and one should decide in advance how many and which meetings one is going to attend. The question of how many meetings to attend is usually determined by one's employer. Most organizations send their scientific employees to 1–3 meetings, depending on their seniority. But how may one select the meeting(s) in which one would like to participate?

4.5.1. World Meetings

The best source of information about forthcoming scientific meetings is *World Meetings*. This is a quarterly publication, published in two parts: *World Meetings outside the United States and Canada* and *World Meetings in the United States and Canada*, giving information about scientific meetings up to two years in advance. Each of the two parts consists of a detailed listing of meetings followed by various indexes and a directory of sponsoring organizations. The listing of the meetings (main entry) is composed of eight subsections, one for each quarter of the two-year coverage period. Each new issue omits the first subsection of the preceding issue while adding a new subsection. Each issue is completely revised, additional meetings being added to all sections and new information being added to old entries.

In the main entry section are given registry numbers to the various meetings. A registry number is assigned to a meeting when it first enters into the *World Meetings* data base and it is the permanent designation for the meeting. The registry number should be used to follow a meeting in which one is interested from one issue to another in order to check for any changes. The registry number consists of seven digits: the first two designate the year in which the meeting will be held, the third indicates the quarter of the year, and the last four digits are serially assigned numbers which were given when the meeting was entered into the data base: e.g. registry number 8720045 indicated that the meeting is scheduled for the second quarter of 1987 and its number is 0045 within that quarter. Following the registry number are the full title of the meeting, its location, its inclusive dates, and the following information related to the meeting: a list of all sponsors of the meeting, name(s) of person(s) or organization(s) to contact in order to obtain additional information, a description, or a list of the various sections and/or subjects of the meeting (this may contain information about the number of papers to be presented, languages used, etc.), estimated number of participants and any restrictions or limitations on attendance, deadline for submission of abstracts and/or papers, information about availability of abstracts and/or proceedings, and information about exhibits connected with the meeting.

The main entry section is arranged in ascending order of registry numbers. The meeting listed in the main section can be retrieved by the use of one of the five indexes—Keyword Index (each meeting may have up to five keyword entries derived from its title or content), Location Index, Date Index, Sponsor Index, Deadline Index.

The Sponsor Index also serves as the Sponsor Directory, giving all the information available concerning the sponsoring organization(s), its name, acronym, original language name (if it is other than English), contact's name, the title of the contact within the sponsoring organization, sponsor's address, cross-references to other sponsor entries.

The value of *World Meetings* is not only that it informs the scientist about forthcoming meetings and helps him to choose which meeting(s) to attend, but also that it gives information about the availability of abstracts and proceedings. This information enables interested workers who are unable to attend a specific meeting to obtain its printed products (abstracts and/or proceedings).

4.6. Information on research in progress

All of the current awareness devices discussed so far tell us about work which reached a 'reported' stage a few months prior to publication. Scientific papers published in journals are the result of projects which had been going on for some time and reach completion (in whole or in part) a few months earlier. The heavy dependence on secondary services in one's current awareness programme as sources of current literature references usually makes these works less current. How can one obtain information on the existence of a research project as early as possible?

The best and easiest way is personal communication. However, as has been pointed out earlier, the size of the scientific community today is such that oral or written communication with all or even most of the people working in one's own and related fields is impossible. Furthermore, science has become so competitive that many investigators are reluctant to give or exchange detailed results of ongoing research. However, there is a 'non-conventional' literature which gives information on research in progress. Examples of 'non-conventional' literature, also known as 'grey literature', are scientific working papers and local authority documents.

4.6.1. The National Technical Information Service

All contractors and grantees of the US Federal and Local agencies are required by their sponsors to submit a research progress report from time to time, followed by a final report at the end of the contract or grant period. One can learn about the development of a project in which one is interested from these technical reports. How can one learn about these reports and obtain them? Summaries of unclassified reports of research sponsored by US Federal and/or local governmental agencies as well as some other specialized information (e.g. analyses prepared by US Federal and local agencies, special technical groups, or their contractors) are reported in one or more of the 26 different weekly publications published by the National Technical Information Service (NTIS)—*NTIS Abstracts Newsletter*. Of the 26 subject categories, 6 are of interest to chemists (Agriculture and Food, Chemistry, Energy, Environmental Pollution and Control, Library and Information Sciences, Medicine and Biology). A biweekly cumulation of all newsletters (*Government Reports; Announcements and Index*) is published for the use of libraries, technical information specialists, and those interested in having all summaries in one volume. The journal is divided into two parts: the main entry and the index. The main entry contains all the abstracts of the various reports arranged according to field and group (all entries are divided into 22 fields). They are arranged alphabetically within the group according to their accession number (NTIS ordering number). There are five different indexes—Subject Index, Personal Author Index, Corporate Author Index, Contract Grant Number Index, and Accession Report Number Index—titles being included in all of them except the Contract Grant Number Index.

Desired scientific reports located through the NTIS publications can be ordered for a modest fee. Orders should be placed with the US National Technical Information Service, 5285 Port Royal Rd., Spingfield, VA 22161, USA. Research reports can be obtained by using the NTIS *Selected Research In Microfiche (SRIM)*. *SRIM* provides its subscribers with full research reports selected for them according to their requirements (e.g. specific subjects, certain investigators).

Information about ongoing US federally funded research can be found in the *Federal Research in Progress (FEDRIP)* which is available online only. It provides access to information such as project title, keywords, starting date, estimated completion date, principal investigator(s), etc.. It is produced by the NTIS and continues, in part, the functions of the now defunct Smithsonian Scientific Information Exchange (SSIE).

4.6.2. European Grey Literature

Information about grey literature originated in Europe (mainly in the European Communities countries) could be found in the *System for Information on Grey Literature in Europe (SIGLE)*. This information is available in a computerized form only. It is a joint production of a consortium of European documentation centres [e.g. the BLDSC, the Saday Nuclear Research Centre, German Information Centre for Energy, Physics and Mathematics (FIZ)].

4.6.3. Programmes and proceedings of meetings

Many research results are reported at various scientific conferences and meetings, often 12–18 months prior to their publication in the scientific literature. Are there any tools to enable us to find out what has happened at the scientific meetings without being present there?

The ideal solution is to obtain the meeting programme. The programme of a meeting is sometimes available only concurrently with the meeting itself. How can one obtain the programme of a meeting in which one is interested? One way is to contact the organizers of the meeting and ask them for a programme. [All the information necessary for this task can be located using *World Meetings* (see Section 4.5.1).] Another way is to use *Conference Papers Index*.

4.6.3.1. Conference Papers Index

Conference Papers Index (formerly *Current Programmes*) is a collection of lists of papers presented at most of the important scientific meetings. It is a monthly publication with cumulative quarterly and annual indexes. Each issue is divided into 17 sections, of which the sections on Biochemistry, Chemistry, and Chemical Engineering are of interest to the chemist. Each section contains detailed information about every meeting covered, including title, *World Meetings* registry number, dates, location, sponsoring organizations, information on ordering of abstracts and proceedings, and a list of papers presented which gives the title of each paper followed by the first author's name and address if available. Each paper has a citation number assigned sequentially throughout the year. The Issue Index contains four indexes: Subject Index, Author Index, Index by Date of the Meeting, and Index by *World Meetings* Registry Number. The last index enables the searcher to connect his search with information obtained some time ago by using *World Meetings*.

Once the desired papers presented at a meeting have been located, one can approach the author for an abstract or a preprint. Furthermore, the correspondence at that stage could be developed into an information exchange or even collaboration.

Information obtained by the use of *Conference Papers Index* could be used at a later stage to verify the publication of the desired proceedings with the aid of *ISTP* (see Section 3.4.7).

4.7. Reviews

Although review articles are very useful as an entry to retrospective searches, they also have a role in one's current awareness programme. One should make a habit of reading review articles covering the latest developments in one's own field during the last year or two. Such reviews are usually published by various serial publications or in review journals. Examples of review serials are *Specialist Periodical Reports* published by the Royal Society of Chemistry (Inorganic Biochemistry; Amino Acids, Peptides, Proteins; Organophosphorus Chemistry; Inorganic Reaction Mechanism; Organometallic Chemistry; and Electron Spin Resonance are just a few of the topics covered) and *Annual Reviews* published by Annual Reviews Inc., Palo Alto, California, USA (*Annual Reviews of Biochemistry, Annual Reviews of Physical Chemistry, Annual Reviews of Energy,* etc.). Reading reviews of this kind enables the reader to obtain a general critical overview of the present development in his own area of research as well as to examine the effectiveness of his current awareness programme.

One may locate the relevant review articles by using the various secondary literature services products discussed earlier (e.g. in the *CA* printed indexes reviews are designated by adding R prior to the abstract number) and/or by

68

using the specially designed CAS and ISI publications and products for that purpose.

4.7.1. CA Review Index

CA Review Index (CARI), which has been published since 1975, is an index to all review articles reported in CA. It is published twice a year (each issue corresponds to a CA volume) under agreement with CAS by the United Kingdon Chemical Information Service.

4.7.2. Indexes of reviews in chemistry

Since 1971 the Royal Society of Chemistry has been publishing the Index of Reviews in Organic Chemistry (IROC) and since 1983 the Index of Reviews on Analytical Chemistry (IRAC). Those publications provide information on review articles in organic chemistry or analytical chemistry, respectively, published in English, German, or French. Contrary to CARI, the IROC and IRAC provide a brief abstract if the scope of the review is not clear from its title.

4.7.3. Index to Scientific Reviews

Index to Scientific Reviews (ISR) is published semiannually by ISI. Each issue includes references to review and 'review-type' articles published during the covered time period. It contains all items published in the review journals and monograph review series which are covered by the SCI and 'review-type' articles from the journals covered by the SCI. These 'review-type' articles are chosen according to the following criteria: articles with words in their title indicating a review article (e.g. review, progress, advance), articles with 40 or more references, or articles coded R (review or bibliography) in the SCI data base. Some of them (those with less than 70 references) are carefully checked for whether they should be included in the ISR.

4.7.4. ISI Atlas of Science

The ISI Atlas of Science provides a concise and factual representation of subjects which are in the active research fronts. Two prototypes of the Atlas have been published, the ISI Atlas of Science, Biochemistry and Molecular Biology 1978/1980 was published at the end of 1981 and the ISI Atlas of Science, Biotechnology and Molecular Genetics 1981/1982 was published at the begining of 1985. Each Atlas covers over a hundred subjects and each subject (or chapter) consists of an essay or a minireview (of about 750 words long) tracing the historical development of the subject, a cluster map [the cluster techniques, their usage, and applications are described and discussed by E. Garfield in Current Contents—(40)5 (1980); (41)5 (1980)], bibliography of the core documents, and a list of current articles which cite those core documents.

The current plan of the ISI is to create at least 5000 minireviews each year for a large number of research fronts, many of them in chemistry.

4.8. Books

One can learn about the publication of new books from the various publications and/or services provided by CAS or ISI. Another source that can be used is *Chem Book*. It is a subject classified bibliography of new and newly announced books in chemistry (written in English, German, French, Italian, and Spanish), published by Karger Libri, Basel, Switzerland.

However, with the large number of scientific books published these days, how can one learn something about a particular book so that one might be able to recommend it to one's students (as a textbook or as a compulsory or suggested reading material), to one's institutional librarian (for purchasing), or to a colleague?

The best means of evaluating a book is to read the published book reviews. Reviews of chemistry books are published by various scientific journals (e.g. *Nature*, *Science*, *Journal of the American Chemical Society*, *Journal of Chemical Education*). However, one has to keep in mind that the average book review is published 6–12 months after the publication of the book itself. In any case, owing to the time taken to write and publish a book, many books are already out of date to some extent (large or small) the moment they are published. In order to make book reviews as useful as possible to the scientific community, journals should attempt to publish them rapidly. This could be done by choosing reviewers willing to give the highest priority to writing reviews, and expediting publication as soon as the reviews have reached the editorial office. Another method is to send book proofs by publishers to journals for reviewing before final publication.

4.8.1. Book review indexes

Book reviews are not covered by any of the abstracting and indexing services. There are some well known book review indexes such as *The Book Review Digest* and *Current Book Review Citations*, both published by H.W. Wilson Co., New York, N.Y., and the *Book Review Index* published by Gale Research Co., Detroit, Michigan. Unfortunately their coverage of scientific books is very poor. The Special Library Association publishes the *Technical Book Review Index*, which covers a very small number of books and has a lag time of 6–10 months from publication of the review to the time of its indexing.

During the years 1980–81 ISI published *Index to Book Reviews in the Sciences* (*IBRS*). The reviews reported were taken from the journals in the SCI data base as well as from publications devoted mainly to book reviews (e.g. *New York Times Book Review*). Unfortunately, because of its failure to achieve financial independence (it had reached a circulation of only 350 copies, which was not able to support its publishing costs) ISI decided by the end of 1981 to stop its publication.

4.9. Computerized services

To go through, scan, and evaluate all of the current data one is interested in, by using the various sources described above, would take much more than the 5–10 hours per week which should be devoted to one's current awareness programme. Further, the ever growing amount of published material will make this task more difficult each year. A solution to the problem is to use computer search services in one's current awareness programme.

4.9.1. Selective dissemination of information

The way to bring to a minimum the amount of material one has to go through in one's current awareness programme is to use computer-readable data base(s) and let the computer itself do most of the scanning. This is known as selective dissemination of information (SDI), a technique used for screening the current literature and notifying the interested party (the user) on specific publications pertinent to his current interests. The information needs of the user are expressed in terms of descriptive terms known as a profile. The profile is compared with the newly acquired information, and whenever there is a match between a source and the profile (a 'hit') this source is sent to the user.

A more general approach to SDI has already been discussed earlier — the standard interest profile (or macroprofile). In such cases the profile is drawn up by outsiders in order to meet the common interests of a large number of chemists. There is no question that the custom-tailored SDI approach yields much better results than the group approach. Here the profile used is a personal profile designed by the chemist himself, with the help of the information scientist, in order to meet his own needs, and could be changed easily whenever there is a need.

The most widely used personal computerized current awareness programmes for chemists, are those based on the CAS or the ISI data bases obtained weekly or biweekly as computer printout in the mail and/or by electronic mail. One can obtain those SDIs directly from the data base producer [e.g. *CA* Alerting Services from CAS, ASCA (Automatic Subject Citation Alert) or ANSA (Automatic New Structures Alert) from ISI], from the vendors of computerized search systems (e.g. BRS, Dialog, Orbit, STN International), or from an information centre in one's own country or even in one's own instutution.

National information centres provide a wide variety of current awareness and retrospective searches on a commercial basis. Their services are based on leased tapes from data base producers (e.g., CAS, ISI, NTIS) as well as on online searching of the various data bases themselves.

It is preferable to use the services of an information centre rather than those of the data producer for the following reasons: the information centre is located in one's own country (sometimes in one's own institution), thus enabling one to have better and faster communication with the information scientist during the preparation of the profile as well as later on; the information centre has access to more than one data base or data bank so that the information scientist could suggest the best data base to be used and could recommend a change in the data base used if the interests of the user changes or if there are changes in the various data bases themselves.

4.10. Keeping one's records and files

An essential part of a good current awareness programme is a good recording and storage system—a system which follows up the journals one reads, telling one which issues of the journals included in the programme one has already read, and a system which enables one to retrieve easily relevant information collected in the programme yesterday, a few months ago, or even years ago.

One should list all the primary and secondary sources in the current awareness programme. The list should include all the primary journals, all the abstracts, index and/or title journals, and all the information and documentation sources one uses. The list could be arranged in a hard-backed notebook, devoting a page to each journal or service, or filing cards could be used, having a separate card for each of the items in the programme. One can use an 'electronic directory' or 'electronic filing' system on a microcomputer for that purpose.

After reading or scanning through a journal one should write the following information on the journal page or card (the journal entry): volume and issue numbers, date of publication, and date of reading. When a service has been used one should write the coverage period of the service used and the date of usage. Checking the list from time to time enables the chemist to notice missing issues or service periods in his programme so that he can attempt to obtain the missing data.

One can locate a relevant scientific article either directly from the primary literature or with the aid of the secondary literature (abstract journals, information services, etc.). After locating an article of interest in the literature one can photocopy the article from the original journal for one's personal usage, write to the senior author for a reprint, or use one of the various services for obtaining a copy of the original article (e.g. Genuine Article of ISI, CASDDS, various commercial services) (copies can be ordered from any of the services by mail, phone, telex, or electronic mail).

It is recommended that one should have a bibliographic list of all articles of interest which have been located via the primary or secondary sources. Titles should be removed from the list once a reprint or a photocopy of the original work has been obtained and transferred to one's permanent files. This list will enable the chemist to follow up articles which are of interest to him before their entry into his files.

In one's current awareness programme one locates data not only in the form of scientific articles but also in other forms (e.g. information about scientific progress, trends in scientific thoughts, conference programmes and abstracts, information about projects in progress, personal communication). All of this relevant information should be transferred into the corresponding categories in one's files. Any information obtained which should be followed up (e.g. projects in progress, book announcements, conference programmes) should be put on a special list for that purpose.

All reprints, data, information sheets, etc., should be filed in a way that will permit their easy retrieval whenever they are needed.

There is no simple easy solution for an effective personal filing system. People usually develop their own systems and some are better than others. The most common system (which unfortunately is a less effective one) is to file the reprints alphabetically by authors. Another system is filing according to subjects; the weakness of this approach is that a reprint is a physical thing and can be filed in only one place, and therefore can have only one subject entry. In spite of the weakness of these two systems they are widely used and some of their users are able to easily locate any desired information they are after. However, many users have to rely on the published indexes which are in the library in order to find the entry to the information in their own personal files. In such cases one may wonder why a reprint file should be built and maintained.

The best solution is to have a personal computerized information storage and retrieval system. This system could be based on a personal microcomputer, the library computer, or one's institution computer. The software of the system is the key for a good and effective retrieval programme. The software includes, in addition to programmes needed to operate the computer, also an effective indexing system. To each paper, data sheet, or any other piece of information a serial number should be assigned (the best is a running number having four digits, but any other number of digits or combination of digits and letters could be used). One should feed the computer with the serial number, the complete bibliographic information, the various indexing terms used (the maximum number is determined according to the hardware and software specifications, but eight to ten indexing terms per item should be enough), and even an abstract of the document. Information can be retrieved by obtaining references for documents filed under specific terms. By doing so one would obtain, in addition to the complete bibliographic data of the documents, their serial numbers which will enable him to locate them easily in his own files.

Microcomputers could be used to obtain (online or by electronic mail) customerized personal information retrieval (e.g. SDIs). Various softwares are available for such purposes—e.g. STAR, B-I-T-S, BRS Search, PULSAR,

Sci-Mate, In-Search, Communicate, KTALK, Dialog Link, STN Connector. Some of these programmes could be used for downloading and processing the current awareness data directly into one's own personal information system.

CHAPTER 5

How to Conduct a Literature Search

One of the questions asked most often by young chemists is how to conduct an effective literature search. There is no simple answer to this question, but the few suggestions and guidelines given below might help the reader to obtain all the desired information he is after via the shortest most effective route.

(a) Define definite goals and specific objectives. What exactly are you looking for? Why do you want it? What will you do with it once you have obtained it? Are you interested in the subject in general or in a specific aspect? (For example, are you interested in the general aspects of cyclohexane dehydrogenation? Perhaps you are looking for something specific—dehydrogenation of cyclohexane by platinum catalysis? Or perhaps what you really want is information about the effect of the carrier and the platinum concentration on the dehydrogenation rate of cyclohexane?)

(b) Before starting your search, find out if you already have any relevant information and, if so, see what leads you may obtain from it. Talk with people around you and see if any of them have any information or leads which may help you.

(c) Decide how soon you need this information—a day, a week, or maybe a month (remember that it takes time to obtain material which is not held by your library).

(d) Decide which time period should be covered by your search—the current year, the last five years, or maybe a longer period (e.g. if you are interested in the crystal structure of XeF_2, then knowing that rare gas compounds were first discovered in 1962 there is no use covering the period prior to that date).

(e) Decide whether to include or exclude certain types of sources (e.g. theses, patents, governmental reports) in your search.

(f) Decide which sources are going to be most fruitful, how readily available they are, and whether you know how to use them. Decide whether to start your search by using a general source such as the *CA* and its indexes, or to examine a more selective source such as *Beilstein's Handbuch der Organischen Chemie* or *Gmelin's Handbuch der Anorganischen Chemie*. Is your problem a very specific one? Perhaps the best results could be obtained by starting with a small, highly specialized source (e.g. if you are looking for information on the usage of CD or ORD for determination of configuration then you should first consult H.B. Kagan, *Stereochemistry*, Vol. 3, *Determination of Configuration by Dipole Moment, CD or ORD*, Georg Thieme Verlag, Stuttgart, 1979).

(g) Decide which part of your search should be done manually and which one online.

(h) Decide which index (or search field) should be used to enable you to obtain the best results using the source of your choice (e.g. working manually with the *CA* best results are obtained using the Chemical Substances and the General Subjects Indexes (prior to 1972 the Subjects Index) and not the Formula Index. However, there are cases in which one should use one or more of the other indexes—Author Index, Patent Index, etc.

(i) Look backwards with a smile. Once you have decided upon a source start with its latest issue, volume, or edition. Use cumulative or collective volumes and/or indexes whenever they are available. For example, when searching the *CA* manually start with the latest available volume index and work backwards. Use the single volumes indexes until you reach the period covered by the Collective Indexes, then use the Collective Indexes in retrospective order.)

(j) Design your search plan beforehand—decide which sources you will use first and which ones you will search at a later stage. Develop your search profile (subjects, keywords, chemical compounds, etc.) before starting; however you should be open to adding or deleting items to or from your profile during the search.

(k) Keep a systematic record of your progress (e.g. sources used, index terms, information found, etc.).

(l) Record your findings in your files.

(m) Be alert to any other pertinent information connected with your other interests. Whenever you locate such information, record it in your files for future use.

(n) Know when to stop your search.

CHAPTER 6

Obtaining Numerical Data

Obtaining reliable, critically evaluated numerical data is a problem concerning all scientists in whatever field they may be active (e.g. biosynthesis, chemical engineering, earthquake detection, environmental quality, IR and/or UV spectroscopy, nutrition, solar energy, thermodynamics). Chemists are interested mainly in numerical data on the physical properties of chemical compounds and/or chemical systems. Many scientists find it difficult to locate the evaluated, or even the raw, numerical data they need. However, all efforts should be made to obtain the best available data, as incorrect or incompletely understood data may lead to erroneous conclusions which could result in systems not working as expected, wrong decision making, etc.

One has to keep in mind that there is a large number of data sources and data banks, and even more are being compiled at present. How can one be aware of their existence?

The formation and compilation of numerical data sources and data banks is a task which is undertaken by various international and national organizations, and knowledge of such organizations will help the scientist in locating the numerical information that is required.

6.1. International and national data collecting organizations

6.1.1. International organizations

The importance of preparing and distributing reliable numerical data for chemists and physicists was realized by IUPAC immediately after its establishment. One of the first decisions taken by this organization (even prior to the approval of its by-laws) in the early part of 1919 was to produce and publish the *International Critical Tables of Numerical Data, Physics, Chemistry and Technology* (which are known today as the *International Critical Tables*, or *ICT*).

The compilation and publication of the *ICT* was undertaken in close collaboration and cooperation with the International Research Council (IRC). This was the beginning of a long and fruitful cooperation between IUPAC and IRC, and later with its successor the International Council of Scientific Unions (ICSU). These tables were published in eight volumes under the editorship of E.W. Washburn during the years 1926–30. Although it has been out of print for many years, no additional edition of the work has ever been published and the compendium has been superceded by new collections. However, even today one may be able to find in the *ICT* some valid data that is not likely to be found easily in other sources. When using the *ICT* one should be aware that many of the numerical and constant values differ from the currently accepted ones (e.g. atomic weights).

The predecessor of IUPAC, the International Association of Chemical Societies, was one of the sponsors of the *Tables Annuelles de Constantes et Donnees Numeriques (Tables of Constants and Numerical Data)* at the begining of the century. The *Tables Internationales de Constantes Donnees Numeriques—Constantes Selectionees (International Tables of Selected Constants)*, which were published in the period 1947–70, are successors to the above tables. They consist of a series of 17 volumes (the first five being in French and the remainder in English and French), each volume being a collection of tables on a specific topic which was of current interest at the time of publication (e.g. oxidation–reduction potentials, spectroscopic data of diatomic molecules).

IUPAC today is very active in collecting and compiling numerical data sources for chemists. Full details about ongoing projects can be found in the latest edition of *IUPAC Current Programmes*, which is published every two years. IUPAC publications include the *Chemical Data Series* (e.g. *International Thermodynamic Tables of the Fluid State*—a collection of monographs each devoted to one element or chemical compound, *Stability Constants*, *Dissociation Constants*) and the *Solubility Data Series*. All IUPAC publications are published by Blackwell Scientific Publications, and an up-to-date list of the whole IUPAC publication programme can be found in the latest Blackwell Scientific Catalogue.

Whereas years ago it was possible for one organization to compile all existing numerical data of the physical sciences (e.g. *ICT*), it is now an impossible task (the size of compendium containing only up-to-date data given in the *ICT* without any indexes would be over 200 volumes of 500 pages each). However, many organizations have undertaken to collect, evaluate, and publish numerical data in a specific area or on a specific subject. IUPAC activities which have been described above are an example of one such organization activity.

In order to coordinate and direct the collection of and to improve the quality, reliability, and accessibility of numerical data on an interdisciplinary basis, ICSU established in 1966 the Committee on Data for Science and Technology (CODATA). CODATA is governed by the General Assembly, which is composed of delegates of member countries and representatives of the various ICSU member unions. CODATA scientific endeavours are achieved via the work of task groups, special *ad hoc* working groups and various panels. Open discussions on the work of the various task groups, special groups, and panels are carried on at the biannual CODATA Conferences. This is also the place where general topics of relevance to CODATA ongoing programmes are presented and discussed.

The CODATA publication programme is composed of the *CODATA Newsletter*, *CODATA Bulletin*, and *CODATA Special Reports*. The *Newsletter*, which is published a few times a year, describes the various activities in the fields of data collection, evaluation, and assessment carried out by CODATA, CODATA member countries, and ICSU member unions. It also lists new publications dealing with data and/or data collection. The *Bulletin*, which is published infrequently (e.g. in 1979 five issues totalling 151 pages were published, in 1980 four issues totalling 312 pages were published, in 1983 two issues totalling 94 pages were published, in 1984 six issues totalling 520 pages were published), deals with the various subjects covered by CODATA (one issue per subject) such as directories of various data sources, abstracts of the CODATA Conferences (the full proceedings from the 4th to the 8th Conference were published by Pergamon Press while those of the 9th Conference onwards are published by North-Holland), and recommended constant values in various areas of research. A complete list of the various *Bulletins* can be found on the back cover of the latest issue. The *Special Reports* are publications which are of interest to limited groups and consist of reports of the various task groups and panels of CODATA.

6.1.2. National organizations

Many data collections and evaluations are carried out by national organizations in many countries. We shall describe below the various activities which are carried out in the USA.

In the USA the collection and evaluation of numerical data is the responsibility of the National Bureau of Standards (NBS), which is part of the US Department of Commerce. In 1963 the National Standard Reference Data System (NSRDS) was established, its activity and policy being affirmed by the Standard Reference Data Act of 1968 (Public Law 90-316). Under this Act the NBS has the sole responsibility for providing critically evaluated data to the scientific community. The whole programme—collection, evaluation and compilation of data—is coordinated by the Office of Standard Reference Data (OSRD). The programme is restricted to well defined physical and chemical properties only (biological properties are not included) of well characterized substances and systems. Large natural systems such as the atmosphere or the oceans are excluded.

The various projects in the programme fall into four different although sometimes connected categories; physical science data, material utilization data, industrial process data, and energy and environmental data. Those projects are conducted in centres located in government laboratories (including the NBS itself), industry, and universities. These centres are engaged in the evaluation and compilation of numerical data from the scientific literature. An up-to-date list of all the data centres and projects administered by the OSRD can be found in the latest edition of *Critical Evaluation of Data in the Physical Sciences—A Status Report on the National Standard Reference Data System*, published by the NBS. An address list of all the data centres, those administered by the OSRD as well as affiliated ones, can be found in Appendix II (under the heading Data Collection and Analysis Centres) of the latest edition of the *CA Index Guide*.

The end products of the programme are critically evaluated data compilations, most of them being published by the NSRDS. However, there are instances in which the publication is carried on by the data centre itself. All publications are listed in the latest NSRDS publication list, which includes prices as well as ordering instructions. The publication list, and any information concerning the various NSRDS activities can be obtained from the Office of Standard Reference Data, NBS, Washington, DC 20234, USA. The OSRD is also attempting to answer enquiries concerning data in the physical and engineering sciences; custom services which require a substantial amount of time are charged to the customer.

In addition to the NSRDS data centres, there are independent data centres which are affiliated to the system. The two best known independent centres are the Thermodynamic Research Centre (TRC), which is located at Texas A & M University, College Station, Tex., USA, and the Center for Information and Numerical Data Analysis and Synthesis (CINDAS) which is located at Purdue University, West Lafayette, Ind., USA. The best known products of those centres are the second edition of the *Comprehensive Index of API-44—TRC Selected Data on Thermodynamics and Spectroscopy*, edited by B.J. Zwolinski (published by the TRC) and a multi-volume series entitled *Thermophysical Properties of Matter*, edited by Y.S. Touloukian (published by IFI/Plenum, New York).

The OSRD is cooperating with various international and national organizations as well as with any other interested parties in the area of numerical data. One of the end products of the cooperation between the OSRD and IUPAC is the *Bulletin of Chemical Thermodynamics*, formerly known as the *Bulletin of Thermodynamics and Thermochemistry*. This is an annual publication which provides information about research in progress in the area of chemical thermodynamics, or about research which has been completed but is still unpublished. No actual data are given in the *Bulletin* as it really provides information on who determined particular properties of a compound or a class of compounds over a specified range of variables.

Another kind of cooperation is that between the OSRD, the Gmelin Institute fur Anorganische Chemie, and the Max-Planck Institut fur Metallforschung in the Ternary Alloy Phase Diagram project. It is a compilation and critical evaluation of all the available data on ternary alloy systems. The compilation will be published in a multi-volume series as part of the *Gmelin Handbook*.

A different type of cooperation exists between the OSRD and various scientific societies. The American Chemical Society and the American Physical Society publish jointly the *Journal of Physical and Chemical Data* for the NBS. This is a quarterly publication which provides critically evaluated compilations of physical and chemical properties data. The last issue in each volume (up to volume 9) contains a Data Compilation Abstracts section, which lists and describes very briefly new selected data compilations. There are instances in which a compilation is too large to be included in the journal and in such cases it is published separately as a supplement to the journal (e.g. *Physical and Thermodynamic Properties of Alcohols*, 420 pp; *The Energetics of Gaseous Ions*, 783 pp; *Heat Capacities and Entropies of Organic Compounds in the Condensed Phase*, 288 pp; *Evaluated Kinetic Data for High Temperature Reactions. Volume 4: Homogeneous Gas Phase Reactions of Halogen- and Cyanide-Containing Species*, 722 pp). A cumulative list of all articles published and the various journal supplements is published at the end of each issue of the journal (the same list also appears in the latest NSRDS publication list). The American Society for Metals publishes the *Bulletin of Alloy Phase Diagrams* as part of the joint data programme on alloy phase diagrams with the NBS. It is a quarterly publication containing provisional evaluated phase diagrams, bibliographic citations to recent phase diagram data, and announcements and reviews of phase diagram conferences, book reviews, and various news concerning phase diagram research.

6.2. Handbooks

Usually a handbook is the first place to check for numerical data. Handbook, according to Webster, is a compact reference book on some subject. The handbook most widely used by chemists is the *CRC Handbook of Chemistry and Physics*, edited by R.C. Weast and published by the Chemical Rubber Company (CRC) Press, Boca Raton, Fla., USA (a new edition, usually with only minor changes, is published annually). CRC Press also publishes a large number of other handbooks, some of which are of interest to chemists (see Table 6.1 for a partial list of CRC handbooks).

Handbooks can and should be used to obtain numerical data, e.g. the temperature at which the vapour pressure of isopropyl alcohol (2-propanol) is 5 atm. When using a handbook one should first consult the index and/or table of contents in order to locate the desired table and/or entry. Under the term Vapour Pressure in the *CRC Handbook of Chemistry and Physics* one finds 17 different entries. Two of the entries deal with vapour pressure of organic compounds—Organic compounds greater than one atmosphere, D206–207, and Organic compounds less than one atmosphere, D191–206. If one turns to the first entry, the desired information is reported in the table on page D207—2-propanol has a vapour pressure of 5 atm at 130.2 °C.

A similar problem is to find the dipole moment and the dielectric constant of Br_3P. The index of the *CRC Handbook of Chemistry and Physics* contains 16 entries for dielectric constants. The one that is of interest is Liquid inorganic pure, E55–56. There are also seven entries to dipole moments,

Table 6.1   CRC handbooks in chemistry (partial list)

CRC Handbook of Antibiotic Compounds
CRC Handbook of Biochemistry and Molecular Biology
CRC Handbook of Chemical Synonyms and Trade Names
CRC Handbook of Chemistry and Physics
CRC Handbook of Chromatography
CRC Handbook of Food Additives
CRC Handbook of Laboratory Safety
CRC Handbook of Mass Spectra of Drugs
CRC Handbook of Material Science
CRC Handbook of Organic Analytical Reagents
CRC Handbook of Spectroscopy
CRC Handbook of Tables for Organic Compound Identification
CRC Handbook of Terpenoids
CRC Handbook Series in Inorganic Electrochemistry
CRC Handbook Series in Organic Electrochemistry

two of which are of interest—Inorganic compounds, E66, and Inorganic compounds gas phase, E63. The data on pages E55–56, E63, and E66 reveal that although the dipole moments and dielectric constants of various inorganic compounds are reported, neither the dipole moment nor the dielectric constant of Br_3P is included. Before looking for any other source, one should examine at least one other handbook. A very useful handbook for the chemist is *The Chemists' Companion, A Handbook of Practical Data, Techniques, and References*, edited by A.J. Gordon and R.A. Ford and published by Wiley-Interscience, New York, 1973. This handbook contains, in addition to the usual collection of tabulated data, also data which are not available in other handbooks (e.g. bond lengths, linear free energy relationships, solvents for recrystallization, chemical shifts, and coupling constant correlation tables for NMR). It also contains much practical information for the experimentalist. (Unfortunately, the handbook has not been updated although several printings of the 1973 edition have been printed.) The index of *The Chemists' Companion* contains an entry for dipole moments of solvents (page 2), but no entry to dielectric constants. However, on page 2 there is a table of properties of solvents and common liquids, which includes both the desired dipole moment and the desired dielectric constant. The dipole moment of Br_3P measured in benzene is 1.7 Debye and its dielectric constant is 3.9 at 20 °C.

Once you have located the desired numerical data, record it immediately in your notebook, accompanied by its primary and secondary sources. Transfer these data to your files as soon as possible. Record them on a compound, chemical system, or reaction card (in case your record system is a paper one) or field (in case of a computerized system) and use the same card or field for the addition of more data at a later stage.

When using handbooks, one has to bear in mind that they are secondary sources only and do not claim to provide the latest available data; they are incomplete and sometimes out of date. The data reported in many handbooks are not necessarily the best available data as in most cases they are not subjected to critical evaluation and no detail of the conditions under which they were obtained are given. Conditions could be located in order to evaluate the data if the primary source(s) (reference(s)) to the data is/are given. The indexes are often not detailed enough (as we have seen above), so occasionally relevant data are missed or lost. In spite of all these limitations, chemists should continue to use handbooks as the first stage of gathering some specific information. As it will often occur that not everything can be found in these handbooks, sometimes one has to go a long way in order to find an answer to a question which at first sight appears to be a very simple one (e.g. which of the elements has an absorption spectral line at 7602 $\pm$ 1 cm^{-1}?).

Handbooks act only as a first stage in obtaining the desired data; the fact that no data could be obtained using one source does not mean that the data needed are unavailable. One must check several sources as well as the original literature before concluding that the specific data have never been reported (in many cases the data were reported but got lost in the literature).

6.3. Locating melting points and/or boiling points

6.3.1. Organic compounds

Suppose one is looking for the m.p. and b.p. of cyclopropanecarboxylic acid (**6.1**). Where could the desired data be located? The first place to check should be a handbook or a chemical compounds dictionary.

6.1

The *CRC Handbook of Physics and Chemistry* has a big table entitled Physical Constants of Organic Compounds (it occupies about two-thirds of Section C—Organic Compounds—of the *Handbook*). The table contains physical data (e.g. melting point, boiling point, colour, crystallization forms, optical rotation, lamda-max., log epsilon, density, refractive index), formulae and synonyms for over 14,000 organic compounds. The compounds are arranged in the table alphabetically according to their IUPAC names. There is a reference to the source(s) from which the data were recorded (usually it is the *Beilstein Handbuch der Organischen Chemie*).

Looking under cyclopropanecarboxylic acid (compound number C912 in the table) one finds the desired data: cyclopropanecarboxylic acid has a m.p. of 18–19 °C and b.p. of 182–184 °C at atmospheric pressure.

Recently the CRC published the *CRC Handbook of Data on Organic Compounds*. It is an extension of the table, Physical Constants of Organic Compounds, which appears in the *CRC Handbook*. It gives data on approximately 24,000 organic compounds (the same data which is given in the *CRC Handbook*). The data are followed by tables which group together compounds by their melting point, boiling point, empirical formula, and structural formula. A separate table lists the IR, UV, NMR, and MS reference numbers for major sources of spectroscopic data.

6.3.1.1. Heilbron's Dictionary of Organic Compounds

Another source that can be used is the latest edition of *Heilbron's Dictionary of Organic Compounds* or its online version (which is available on Dialog). The *Dictionary of Organic Compounds* (*DOC*) has been proved during the years to be one of the most useful tools not only for organic chemists but also for many other scientists (e.g. biochemists, pharmacists). The fifth edition of the *DOC* was published in 1982, replacing the fourth edition and its 15 annual supplements. The new edition is composed of seven volumes (five volumes of entries and two index volumes) and contains about 50,000 entries totalling about 150,000 different compounds, a number which covers less than 2% of all known organic compounds. One should keep in mind that within the large number of all known chemical compounds there is some kind of hierarchy of importance. While most compounds are of interest and/or importance to very few scientists (in many cases to one or two only), a few compounds are of interest to a large number of scientists (e.g. starting materials, intermediates, solvents, natural products). This is reflected in the number of times a compound is reported in the literature: about 75% of all chemical compounds are reported only once in the literature, 16% reported twice, only less than 10% are reported more than twice! *DOC* is intended to cover the organic compounds which are of considerable importance and interest not only to organic chemists in particular but also to other chemists and scientists. The compounds reported include fundamental organic compounds having simple structures, compounds which have widespread industrial or commercial use, compounds frequently used in the laboratory as solvents, reagents, or starting material, natural products, and any other compounds that have interesting structural or biological properties. Even so one sometimes wonders what are the criteria or why at all certain entries were included in the *DOC* [e.g. 2-phenanthreneacetic acid (**6.2**) and 9,10-phenanthrenequinone-3-sulphonic acid (**6.3**) are entries to the fifth edition of *DOC*, both compounds have been reported in the literature only once each during the years 1965–85].

Yearly supplements are published to the *DOC*; each supplement contains 3000-4000 entries of which about 1500–2000 are completely new entries while the rest are updated. Each supplement contains a cumulative index to all supplements.

All entries are arranged alphabetically according to the Entry Name. Whenever feasible systematic names (*CA* Index Names or IUPAC Names)

84

6.2

6.3

were used. However trivial names were used for most natural products as well as for some of the other compounds. The names used are those which in the Editor's opinion are most likely to be known and/or to be used by most readers (e.g. tetraphenylmethane is used for compound **6.4** and not its *CA* Index Name—benzene,1,1′,1″,1‴-methanetetraryltetrakis; the *CA* Index Name of compound **6.3** is 3-phenanthrenesulphonic acid, 9,10-dihydro, 9,10-dioxo-).

6.4

Each entry contains concise detailed information which includes synonyms and alternative names (e.g. the *CA* 9th and 8th Collective Index Names, names recommended by the British Pharmaceutical Commission, the British Standards Institution, the United States Approved Name Review Board); structural diagram; CAS Registry Number; molecular formula; molecular weight; physical description, sources, usage and importance; physical properties such as melting point, boiling point, spectral data, solubility, density, optical rotation, refractive index, pK values, etc.; safety information such as hazards, Registry of Toxic Effects of Chemical Substances (RTECS) number; bibliographic references [the content of many of the references are indicated by memonic suffices—synthesis by (synth), mass spectra by (ms), ^{13}C nuclear magnetic resonance by (cmr), hazards by (haz), toxicity by (tox), etc.].

Each of the main entries is numbered, the number consisting of an alphabetical letter followed by five digits. The alphabetical letter represents the first letter of the entry name of the compound, the first digit represents the supplement number [in the case of the fifth edition the number is 0, entries at the 1983 supplement (first supplement) have the number 1, entries in the second supplement have the number 2, and so on according to the supplement number], and the other four digits are the number of the entry name within all the entry names having the same first letter.

The *DOC* has four indexes—Name Index, Formula Index, Heteroatom Index, and CAS Registry Number Index. The Name Index connects all names and all synonyms and derivative names used in the text to their entry compound *DOC* number. The Formula Index is arranged according to the Hill System and connects all entry compound formulae to their *DOC* number. The Heteroatom Index is a variation of the Rotated Formula Index (see p. 55) and contains formulae having atoms other than C, H, N, O or halogens, connecting them to their *DOC* number. The CAS Registry Number Index connects all CAS Registry Numbers reported in the *DOC* to the entry compound *DOC* number.

In 1984 a companion three volume dictionary entitled *Dictionary of Organometallic Compounds* was published. It is very similar in its structure and content to the *DOC*. It covers the area of organometallic chemistry and contains about 14,000 entries. Yearly supplements are being published to the *Dictionary of Organometallic Compounds*. In various stages of preparation are other dictionaries such as *Dictionary of Organophosphorus Compounds*, *Dictionary of Steroids*, *Dictionary of Alkaloids*.

Smaller subset publications of the *DOC* in individual areas of organic chemistry (e.g. *Amino Acids and Proteins Sourcebook, Carbohydrate Sourcebook*) as well as subset publications of the *Dictionary of Organometallic Compounds* (e.g. *Organometallic Compounds of Iron, Organometallic Compounds of Boron, Organometallic Compounds of Silicon*) have been published by Chapman and Hall.

The fifth edition of the *DOC*, as well as the other compound dictionaries published by Chapman and Hall has the advantage that it can be used as the starting point for a more comprehensive search. The references listed at the end of each entry can be used for a further manual search of the *CA* and/or *SCI* while the CAS Registry Numbers provide an easy entry to an online search of the *CA*, or any other database using the CAS Registry Numbers (e.g. MEDLINE).

Using the *DOC* one obtains the same data on cyclopropanecarbonylic acid as in the *CRC Handbook of Chemistry and Physics* (m.p. 18–19 °C, b.p. 182–84 °C). However the entry of cyclopropylcarboxylic acid in the *DOC* also gives data on some of its derivatives (e.g. methyl ester, ethyl ester, amide, acylchloride).

How could one find the m.p. of cyclobutanecarboxamide (**6.5**), a compound which is found not to be included in the *CRC Handbook?* One could use the *DOC* as a second source, remembering that data concerning derivatives is given under the parent compound heading. Compound **6.5** is a derivative of cyclobutanecarboxylic acid (**6.6**), and indeed under the entry to cyclobutanecarboxylic acid one finds the amide m.p. — 152–53 °C.

$CONH_2$ on cyclobutane	CO_2H on cyclobutane
6.5	**6.6**

86

6.3.1.2. Beilstein's Handbuch der Organischen Chemie

A completely different approach is to use as a first source *Beilstein's Handbuch der Organischen Chemie* (or *Gmelin Handbuch der Anorganischen Chemie* when one is looking for data concerning inorganic or organometallic compounds). Indeed, using *Beilstein* the following information was found: cyclopropylcarboxylic acid m.p. 18–19 °C, b.p. 182–84 °C; cyclobutanecarboxamide m.p. 152–53 °C.

Suppose a graduate student is interested in the m.p. of 4-nitrothiophenylacetamide (**6.7**) and 3,4,5-triphenyl-2(5*H*)-furanone (**6.8**). He was not able to locate the desired data using the *CRC Handbook* nor using the *DOC* as compounds **6.7** and **6.8** are not reported in those sources. How could he retrieve the m.p. of the above compounds from *Beilstein*?

6.7	**6.8**

Beilstein's Handbuch der Organischen Chemie (*Beilstein's Handbook of Organic Chemistry*) is a collection of critical reviews of all the material on all known organic compounds. The first edition was compiled and produced by Friedrich Konrad Beilstein in 1883. The compendium included, in addition to all the known data concerning all the organic compounds at that time (about 15,000 compounds), sections dealing with organic analysis and the determination of physical constants. The second and third editions were also compiled by Beilstein himself. The fourth and last edition was produced by the German Chemical Society. After the Second World War the preparation and production of the *Handbook* was transferred to the Beilstein Institute, which is located at Carl Bosch Haus, Varrentrappstrasse 40-42, Frankfurt a/M, GFR.

The fourth edition was published during the years 1918–38 and covered the literature up to the end of 1909. This edition consists of 27 volumes reporting about 140,000 organic compounds. Volumes 28 and 29 are index volumes (General Subject Index and General Formula Index, respectively). Upon completion of the fourth edition, it was decided not to produce any further editions, and instead to issue Supplementary Series (Erganzungswerk), each one covering a definite period and arranged according to the Main Work (Hauptwerk). The First Supplement (EI) covers the period 1910–19, the Second Supplement (EII) covers the period 1920–29, and the Third Supplement (EIII) covers the period 1930–49.

Owing to the vast amount of data collected since 1950, it was decided during the publication of the Third Supplement to start publication of the Fourth Supplement (EIV) which covers the period 1950–59, before finishing the publication of the Third Supplement. It was also decided to combine the Third and Fourth Supplements starting from Volume 17 (the begining of the description of Heterocyclic Compounds). Thus the combined Third and Fourth Supplements (EIII/IV) cover the period 1930–59. The publication of the Fifth Supplement (EV), which covers the period 1960–79, started in 1985 while the publication of the Fourth as well as the combined Third and Fourth Supplements were still in progress. The Fifth Supplement is published in English and its publication started with Volume 17. It is interesting to know that about 80% of the chemical compounds which will be cited in the Fifth Supplement will be reported in the *Beilstein* for the first time.

Two additional volumes to the Main Work were published in 1938: Volume 30, dealing with polyisoprenes (e.g. terpenes, carotenoids) with a literature coverage up to the end of 1934, and Volume 31, dealing with carbohydrate chemistry with a literature coverage up to the end of 1919. No supplements to these two volumes have been published and later material on these subjects has been incorporated into the corresponding volumes of the various supplements according to their structure in accordance with the general rules of the Beilstein System; the polyisoprenes were incorporated into the Third Supplement and the carbohydrates into the Fourth Supplement. The arrangement of the whole work is summarized in Table 6.2.

One should remember that the classification system in each Supplement is the same as in the Main Work. Each Volume of any Supplement contains the same classes and only those classes of compounds as the corresponding Volume of the Main Work. This correlation, together with the growth of organic chemistry, necessitated the subdivision of most Volumes of the Third and Fourth Supplements. Thus Volume 6 of the Third Supplement consists of eight subvolumes totalling 7180 pages. It is interesting that, in contrast, in the First Supplement two or three nominal Volumes were combined together into a single bound book.

As indicated above, the Main Work and the various Supplements cover definite periods. Those various coverage periods are not always kept and the various Supplements are considerably more up-to-date than indicated in Table 6.2. Corrections to reported data based on recent research and new important findings (e.g. constitution, configuration) are included past the end of the coverage period.

The Third Supplement covers the period 1930–49. In the entry for 2-aminofluorene in the Third Supplement (Volume 12, page 3285) there is a reference to N. Campbell and A.F. Temple's work which was published in 1957 [*J. Chem. Soc.*, 207 (1957)]. The reason for the early report of this work is that Campbell and Temple proved that various ring closure reactions of 2-aminofluorenone (e.g. Dobner-Miller reaction, Skraup reaction) result in the formation of the corresponding 1*H*-indeno[2,1-g]quinoline (**6.9**) derivatives. These findings contradict the assumption of G.K. Hughes *et al.* [*J. Proc.*

Table 6.2   The fourth edition of the *Beilstein* and its Supplements

	Abbreviation	Number of volumes (subvolumes)	Coverage period	Colour of label on spine
Main Work	H	27*	31.12.09	Green
First Supplement	EI	17†	1.1.10–31.12.19	Red
Second Supplement	EII	26†	1.1.20–31.12.29	White
Third Supplement	EIII	62‡	1.1.30–31.12.49	Blue
Fourth Supplement	EIV	70‡	1.1.50–31.12.59	Black
Third and Fourth Supplement	EIII/IV	77‡	1.1.30–31.12.59	Blue/black
Fifth Supplement	EV	17§	1.1.60–31.12.79	Red

* Two additional volumes are included in the main work: Volume 30—-Polyisoprenes (literature coverage up to 31.12.34) and Volume 31—Carbohydrate (literature coverage up to 31.12.19).

† There is a General Subject Index (Volume 28 in two parts) and a General Formula Index (Volume 29 in three parts) for Volumes 1–27 of the Main Work and the first two Supplements.

‡ Cumulative Indexes to the Main Work and the first four Supplements have been published (Subject Index 1, 2–3, 4, 5, 6, 7–8, 9–11, 17–18, 19, 20–22, 23–25, 26) as additional volumes.

§ Published by the end of 1986.

Soc. N.S.W., **71**, 449 (1938); *CA*, **33**, 609 (1939)], who reported on the same page that these reactions yield the isomeric 1*H*-indeno[2,1-f]quinoline (**6.10**) derivatives.

6.9   6.10

Beilstein's Main Work and the first four Supplements are written in German, but this should not scare anybody not familiar with this language as the vocabulary is simple, limited, and can be learned very easily. For the benefit of users who have a language barrier, the publishers of the *Beilstein* have published special German–English and German–French *Beilstein Dictionaries*. These dictionaries contain about 2000 words—nearly every word used in the *Handbook*. The dictionaries can be obtained free of charge from the publisher (Springer-Verlag New York Inc., 175 Fifth Avenue, New York, NY 10010, USA, or Springer-Verlag GmbH Heidelberger Platz 3, D-1000 Berlin 33, GFR).

One should remember that up to the Fifth Supplement unconventional journal abbreviations were used in the *Beilstein* (e.g. *B.* stands for *Chemische Berichte*; *Am. Soc.* for *Journal of the American Chemical Society*; *Helv.* for *Helvetica Chimica Acta*; *Soc.* for *Journal of the Chemical Society*). The user should use the list of journal title abbreviations which is given in the preface of each volume. In case of any doubt about the meaning of the abbreviation the list should be consulted immediately.

A complete general index to the *Beilstein* is available only for those series which have been completed. Thus there is a General Subject Index (name index of all compounds dealt with) in Volume 28 (which was published in two subvolumes) and a General Formula Index (arranged according to the Hill System) in Volume 29 (which was published in three subvolumes) covering the Main Work and the first two Supplements.

Cumulative subject and formula indexes for individual volumes or groups of volumes covering the Main Work and the first four Supplements are available (e.g. Volume 1; Volume 19; Volumes 2 and 3; Volumes 17 and 18; Volumes 20, 21, and 22). Plans are calling for the production and publication of one cumulative index to the Main Work and first four Supplements; such an index will be covering the whole literature of organic chemistry from its beginning up to the end of 1959.

In addition to all these cumulative indexes each of the volumes of the Main Work and the first two Supplements still has its own subject index. Starting from the Third Supplement each of the subvolumes has its own subject and formula indexes.

Beilstein contains entries for all carbon compounds described in the scientific literature. Each entry contains the following information: structure (including conformation), occurrence (e.g. isolation from natural products, synthesis, purification), physical properties, descriptive properties (e.g. melting point, boiling point, dipole moment, reflective index), structural properties (such as crystal data, bond lengths and angles, conformation, energy data, polarizability), spectroscopic information, physical properties of mixtures of the compound, chemical properties, physiological properties, usage, characterization and analysis, and salts and addition compounds (e.g. hydrates) of the compound. All of this information is collected, evaluated, checked, and double-checked with the greatest possible care for its soundness, internal consistency, and significance. Much of the cited information is followed by critical comments and cross-references. In many cases the information is given in such a way that the user obtains all that he needs just by reading the entry, thus avoiding the use of the primary literature. If a compound sought for is not reported in the *Beilstein*, one can be sure that it was not known by the end of the coverage period of the Supplement used.

Each entry starts with a heading, which contains the most important name of the compound followed by other names and empirical and structural formulae. A back reference is given immediately afterwards. Thus H 99, EI 347, EII 593, EIII 194, which follows the entry to methyldiethylamine in Volume 4 of the Fourth Supplement, means that the previous entries to that compound can be found in the same volume (Volume 4) of the Main Work on page 99, the First Supplement on page 347, the Second Supplement on page 593, and the Third Supplement on page 194. If back references are lacking it means that the compound is described in the Supplement concerned for the first time.

There is also a coordination reference, which is to be found on top of each odd-numbered page of the compendium; it indicates the page of the Main Work to which the data on the page concerned would have been assigned (the latest Supplements also have coordination references to the Third or Third and Fourth Supplements, respectively). Thus, pages in different supplements having the same coordination reference contain information about the same compounds or compounds of similar constitution.

Although the main users of the *Beilstein* are organic chemists, it has proved to be of value also to physicists, physical chemists, analytical chemists, chemical engineers, and biochemists.

Organic chemists prefer to use the *Beilstein* for retrospective search concerning a specified organic compound rather than any other source because of the ease and speed of obtaining the desired information. Suppose a chemist (not necessarily an organic chemist) is interested in all the information he can get on diethylmethylamine. One route is to search for the desired information in the abstract literature. The great disadvantage of this approach is that the information required may be scattered throughout a publication period of over 100 years and in order to cover this period one has to use more than one abstracts journal.

Two abstracts journals cover the literature that was published during the second half of the 19th century and the beginning of the 20th century: *Chemisches Zentralblatt* and *British Abstracts*.

Chemisches Zentralblatt was started in 1830 as *Pharmaceutisches Centralblatt* and underwent a number of changes in its title and publisher (a common phenomenon in that period) until it received its present name in 1897. Its publication lapsed in 1945 and although it was re-established a few years later, it never recovered completely and was discontinued in 1969. *Chemisches Zentralblatt* published cumulative quinquennial indexes since 1925; previously indexes were published every three or four years. Formula Indexes were introduced only in 1922. Using those indexes one should remember that up to 1956 they were arranged according to the Richter System; only in 1956 did *Chemisches Zentralblatt* started to use the Hill System.

In the Richter System (the System was used in the Formula Index of the *Beilstein* Main Work and in the combined Index of the Main Work and the First Supplement), organic compounds are arranged in order of increasing number of carbon atoms in the molecule (C_1, C_2, C_3, etc.) followed by subdivision according to the number of elements present (e.g. $C_6H_6-C_6II$; $C_6H_6O-C_6III$). The atoms of elements other than carbon are listed in the order H, O, N, Cl, Br, I, F, S, P, with any other element listed alphabetically. Thus using the Richter System chlorobenzene (C_6H_5Cl) would appear before bromobenzene (C_6H_5Br) which would appear before m-dinitrobenzene ($C_6H_4O_4N_2$). It is worth while remembering that the order of appearance of the above compounds in the Hill System is completely the reverse—m-dinitrobenzene ($C_6H_4N_2O_4$) before bromobenzene (C_6H_5Br) before chlorobenzene (C_6H_5Cl).

British Abstracts also underwent a number of changes in its title. It started in 1871 as a supplement to the *Journal of the Chemical Society* and in 1882 as a supplement to the *Journal of the Society of Chemical Industry*. In 1926 these two supplements were combined into *British Chemical Abstracts*. In 1939 it became *British Chemical and Physiological Abstracts*, and in 1945 *British Abstracts*. It was terminated in 1953 for financial reasons.

It can easily be seen that going through *CA* for the last 60–65 years followed by 60–70 years of *Chemisches Zentralblatt* and/or *British Abstracts* requires much time and effort. On the other hand, locating the desired information prior to 1960 by using the *Beilstein* is easy and rapid.

In order to locate the desired compound one should use the cumulative index. It is much easier to use the Formula Index than the Subject Index, especially for non-German-speaking chemists because of the vagaries of German chemical nomenclature. Using the Formula Index to the Main Work and the first two Supplements (the latest available cumulative index) one finds under $C_5H_{13}N$ a list of compounds. The first compound on the list is methyldiethylamine— 4 99, I 347, II 593. Thus the entry to our compound is in Volume 4 on page 99 in the Main Work, page 347 in the First Supplement, and page 593 in the Second Supplement. The entries to the Third and Fourth Supplements could be located using the coordination reference. Doing so (the coordination reference is 4 99) one finds the desired entries on page 194 in the Third Supplement, and on pages 321–32 in the Fourth Supplement (remember that all these entries are in Volume 4).

All the needed information up to 1960 is summarized in about a page and a third and was located within a few minutes. The information from 1960 onwards could be located using the *CA*.

Let us go back to our graduate student (p. 86). How would be he able to locate his compounds in the *Beilstein*? If the compounds he is after could not be located using the cumulative Formula Index of the Main Work and the first two Supplements, then they were not reported in the literature up to the end of 1929. What should his next move be? The best way would be to identify in which volume the compound is listed and then to locate the compound using the Volume Index. In order to identify the desired volume one should be familiar with the Beilstein System, a System which determines the place of the various compounds in the compendium. A few basic principles of the Beilstein System are described below.

(a) All organic compounds are divided into three major groups:
   I. Acyclic Compounds which are described in Volumes 1–4.

II. Isocyclic Compounds which are described in Volumes 5–16.
III. Heterocyclic Compounds which are described in Volumes 17–27.
The heterocyclic compounds are further subdivided according to the
type and number of the ring heteroatom (see Table 6.3).

Table 6.3   Main divisions of the *Beilstein* volumes

Compound group:	Acyclic (A)	Isocyclic (B)	Heterocyclic (C)					
			10	20*	1N	2N	3N 4N	Nitrogen and oxygen; other heteroatoms**
Volumes:	1–4	5–16	17–18	19	20–22	23–25	26	27

* Ring heteroatoms may occur in one ring or distributed over several rings. For example,

and both belong to heterocyclic compounds with 20 and

are described in Volume 19.
** E.g. B, Si, and P, but no S, Se, and Te (see rule f.3).

(b) The location of a compound in the Beilstein System is determined by the
structural element (or fragment) that is classified latest in the System—
the Principle of Latest Possible Entry. Thus a compound containing a
heterocyclic ring with one nitrogen, a carbocyclic ring and an aliphatic
chain is classified as a heterocyclic compound irrespective of all its other
structural elements (e.g. compound **6.11** is described in Volume 20 as
a compound with one nitrogen in the heterocyclic ring without any
functional group).

6.11

(c) Classification within a division (Acyclic, Isocyclic) or subdivision (Heterocyclic) is based upon the type and number of the compound functional
groups. One can distinguish between:
I. Registry Compounds
II. Derivatives of Registry Compounds

I. Registry Compounds
The Beilstein System contains 4877 registry compounds. They are
defined as hydrocarbon or heterocycles without any functional
group(s) (they are also known as parent compounds), or with one
or more of the functional groups listed in Table 6.4 and bonded to

a carbon atom. In cases of heterocyclic compounds, the heterocyclic ring must contain at least one carbon atom, the ring heteroatom(s) must be either oxygen or nitrogen (or both of them), and the heteroatom may be neither substituted nor bearing a functional group. The *Beilstein* volume containing an entry for a desired compound may be determined directly from Table 6.4. However if a Registry Compound contains two or more functional groups the Principle of Latest Possible Entry is applied again; e.g. 4 hydroxybenzenesulphonic acid (**6.12**) is classified as an isocyclic sulphonic acid (reported in Volume 11) and not an isocyclic hydroxy compound (reported in Volume 6), benzoic acid (**6.13**) is reported in Volume 9, aniline (**6.14**) in Volume 12, 1,2,diaminobenzene (**6.15**) in Volume 13, while antranilic acid (**6.16**) is reported in Volume 14.

The sequence of appearance within a volume is determined by the degree of saturation. Thus cyclopentane (**6.17**) which belongs to Registry Compound 452, cyclopentene (**6.18**) which belongs to Registry Compound 453, and 1,3-cyclopentadiene (**6.19**) which belongs to Registry Compound 455 are reported in Volume 5 of the Fourth Supplement on pages 14, 209, 377, respectively (5 EIV 14, 5 EIV 209, 5 EIV 377).

6.12   6.13   6.14   6.15

6.16   6.17   6.18   6.19

The number of carbon atoms determine the sequence of appearance within a Registry Compound. Thus within Registry Compound 453 the sequence of appearance is cyclopentene (**6.18**) before cyclohexene (**6.20**), before 1-methylcyclohexene (**6.21**), before 1,2-dimethylcyclohexene (**6.22**) (5 EIV 209, 5 EIV 218, 5 EIV 245, 5 EIV 268, respectively).

All other compounds which cannot be classified as Registry Compounds are Derivatives of Registry Compounds.

Table 6.4 Content of *Beilstein's Handbuch der Organischen Chemie*

Type of registry compound	A	B	Main division					
						C		
			10	20 / 30	1N	2N	3N 4N	Other
Compounds without functional groups	1	5						
Hydroxy compounds (—OH)		6	17		20	23	26	27
Oxo compounds (=O) =O		7			21	24		
=O + —OH		8						
Carboxylin acids (=O / —OH)	2	9						
≡OOH + —OH; ≡OOH + =O; ≡OOH + —OH	3	10						
Sulphinic, sulphonic. selenic, selenoic, and tellurinic acids		11	18	19	22	25		
Amines —NH₂		12						
—NH₂; (—NH₂)ₙ; —NH₂ + —OH		13						
—NH₂ + =O; —NH₂ + ≡OOH	4	14						
Hydroxylamines, dihydroxylamines, and hydrazines		15						
Azo compounds, diazonium compounds, compounds with groups of three or more nitrogens		16						
Compounds containing carbon directly bound to other elements except halogens. O, S, Se, Te, N								

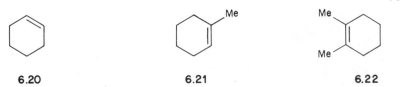

6.20   **6.21**   **6.22**

II. Derivatives of Registry Compounds

These are compounds in which some structural modification should be carried out in order to obtain a Registry Compound.

A. Functional derivatives. These are compounds containing derivatized functional group(s) (e.g. methylacetate). The Registry Compound is obtained by 'formal' hydrolysis, which does not necessarily correspond to a laboratory reaction. The products of the hydrolytic cleavage are such that the OH is added to the first atom from the sequence—C, B, Si, P, S, halogen, N, O. The hydrolysis products may contain only functional groups listed in Table 6.4. One should also remember the Principle of Latest Possible Entry; thus the phenyl-4-pyridylamine (**6.23**) Registry Compound is 4-aminopyridine (**6.24**) (Registry Compound 3393), which is reported in 22 EIII/IV 4098. It is interesting to know that 4-hydroxypyridine (**6.25**), is Registry Compound 3111 and it is reported in 21 EIII/IV 446 (Reaction 6.1).

B. Substitution derivatives, or substitution products. These are compounds having non-functional substituents (F, Cl, Br, I, NO_2, N_3). The Registry Compound is determined by replacing the substituent with a hydrogen atom. However if a carbon atom bears substituent atom(s) and a functional derivative the Registry Compound is obtained by formal hydrolysis of the substituent atom(s) and the functional group; the resulting gendiols are dehydrated to give the corresponding oxo compound as the Registry Compound (Reaction 6.2).

C. Chalcogen derivatives. These are organic sulphur, selenium, and tellurium compounds. The Registry Compound is determined by replacing the bivalent atom with oxygen (sulphoxides, sulphones, and sulphonium compounds are to be found following the corresponding sulphides).

$$\text{C}_6\text{H}_5\overset{\overset{\text{Cl}}{|}}{\underset{\underset{\text{Cl}}{|}}{\text{C}}}-\text{O}\overset{\text{O}}{\text{C}}\text{CH}_3 \longrightarrow \text{C}_6\text{H}_5\text{C(OH)}_3 \longrightarrow \text{C}_6\text{H}_5\text{CO}_2\text{H} \qquad (6.2)$$

$$+ 2\text{HCl} + \text{CH}_3\text{COOH}_3$$

A much more detailed information as well as information about tautomeric and mesomeric compounds in the *Beilstein Handbook* may be found in the booklet '*How to Use Beilstein*' which was published by the Beilstein Institute and distributed free of charge by Springer-Verlag.

A simple and a straightforward procedure for the locating of chemical compounds in the *Beilstein* has been reported in the *Journal of Chemical Education* [J. Sunkel, E. Hoffmann, and R. Luckenbach, *J. Chem. Ed.*, **58**, 982 (1981)]. Recently a software package entitled *SANDRA* (*Structure AND Reference Analyser*) has been developed for IBM-PC and its compatibles. It is designed to enable one to use the *Beilstein* more efficiently. The package specifies the Beilstein System position for any input structure independent of whether the compound actually appears in the *Beilstein*. It is claimed that the output reliability is 99% and the resolution is of the order of 40 pages. For the usage of the infrequent (or casual) *Beilstein* user who is unfamiliar with, and/or afraid of, computer graphics, the Beilstein Key has been developed. It is a computer program that enables the user to locate the volume number containing a particular organic compound by a series of simple questions and answers. A *Beilstein* volume number is generated even if the compound in question has not been reported in the *Beilstein* (because it is not known at all or because it was reported after the closing date of the latest Supplement).

Plans are in progress for an online version of the *Beilstein* which is going to be available to the public at the first stage via Dialog and STN International. The data base is generated from the following sources: the printed *Handbook* from the Basic Series up to the Fourth Supplement (H to EIV), covering the literature from 1830 to 1960; the printed Fifth Supplement of the *Handbook* covering the period 1960 to 1980; some unchecked unevaluated material covering the period 1960 to 1980 (in order to give scientists immediate access to this data); post-1980 material which is now abstracted and recorded electronically.

The data base is going to be divided into two parts, the refined part and the current part. The refined part will contain all the data which had been published in the *Handbook* as well as unpublished data which had been checked and evaluated. The current part will contain unchecked data and abstracts. The refined part will be extended continuously by transferring to it material which has been checked and evaluated from the current part.

Knowing the principles of the Beilstein System our student was able to determine the Registry Compounds which correspond to 4-

nitrothiophenylacetamide (**6.7**) and to 3,4,5-triphenyl-2(5*H*)-furanone (**6.8**), respectively.

Compound **6.7** is a functional derivative, a substitution product as well as a chalcogen derivative. Thus replacing the sulphur atom with oxygen and the nitro group with hydrogen gave the functional derivative phenoxyacetamide (**6.26**); formal hydrolysis of **6.26** gives phenol, glycolic acid (acetic acid, hydroxy), and ammonia (Reaction 6.3).

$$ \longrightarrow \quad + \quad NH_3 \quad + \quad HOCH_2COOH \qquad (6.3) $$

6.26

The Latest Possible Entry is phenol and the desired compound (compound **6.7**) should be reported in the same volume with phenol. The desired volume could be located with the aid of Table 6.4, or by using the *Beilstein* latest Cumulative Index (looking for phenol).

Our student found that phenol is reported in Volume 6 (hydroxy iso-cyclic compounds). Using the Cumulative Formula Index to Volume 6, he found that the desired compound ($C_8H_8NO_3S$) is reported on page EIV 1710. On that page he found the information sought—the melting point of 4-nitrothiphenylacetamide is 141–142 °C.

Compound **6.8** is a Registry Compound and as such should be reported in Volume 17 (heterocyclic compound with one heteroatom, oxygen, having in addition a carbonyl group). Among the many compounds having the formula $C_{22}H_{16}O_2$ our student found in the Cumulative Formula Index to Volumes 17 and 18 an isomer of **6.8**, viz. 3,4,5-triphenyl-2(3*H*)-furanone (**6.27**) (the *Beilstein* name of **6.27** is furan-2-on-3,4,5-triphenyl-3*H*), but not the desired compound. His conclusion was that compound **6.8** had not been reported in the literature up to the end of 1959, and therefore he should turn to the *CA*. The best search tool to use manually in this case is the *CA* Chemical Substance Index (or the *CA* Subject Index for the period prior to 1972). Results could also be obtained using an online search of the *CA* data base.

6.27

6.3.1.3. CA Chemical Substance Index

The *CA* Chemical Substance Index links chemical substances names to the *CA* abstract numbers for the sources in which those compounds are reported. The index is arranged alphabetically using the *CA* Index Names, which are as descriptive as possible and tend to group together structurally related substances. This is done by listing the Index Parent Name first, followed by the substituent name(s) (e.g. 2-bromocyclobutene and 2-methylcyclobutene will be listed under the Index Parent Name—'cyclobutene' followed by an alphabetical order of their substituents names—'2-bromo' and '2-methyl', respectively, 'cyclobutene-2-bromo', 'cyclobutene-2-methyl'). In order to save space and improve readability the Index Parent Name appears only once and is replaced by a long dash for related compounds. When two or more parent names or substituent names differ only in their location, the names with the lower locants appear before those with the higher locants (e.g. the various isomers of undecanedione appear in the following order: 2,5-undecanedione, 2,7-undecanedione, 3,9-undecanedione, and 5,7-undecanedione).

Each *CA* Index Name is followed by a CAS Registry Number (RN) and a description of the content in which the compound is mentioned in the original source. When such a description is missing it is an indication that the emphasis in the source is on its synthesis (it may be a newly prepared compound or a well known one).

Our student realizes that a source which reports the synthesis of compound **6.8** will give its m.p.; therefore he decided to locate a reference describing the synthesis of **6.8**. Before using the *CA* Chemical Substance Index one has to determine the *CA* Index Name of the compound. Looking at 3,4,5-triphenyl-2(5*H*)-furanone our student realized immediately that the parent compound is 2(5*H*)-furanone and the substituents are 3,4,5-triphenyl- (in cases of complex structures the determination of the *CA* Index Name may not be so simple).

Using the *CA* Chemical Substance Index of Volumes 104 (the first half of 1986) and 103 and 102 (1985), no entry for the preparation of 2(5*H*)-furanone, 3,4,5-triphenyl- was found (in the Volume 104 Index there is an entry to the reaction of compound **6.8** with potassium superoxide). In the Index of Volume 101 the following entry was found:— 2(5*H*)-furanone, 3,4,5-triphenyl-[7404-46-8] prepn. and photolysis of, 229794p. Thus the preparation of the desired compound was described in a source that was abstracted in the *CA* and the abstract number is 101 : 229794p (abstract 229794p in Volume 101). On examining the abstract our student found a bibliographic citation to the Lohray *et al.* paper in which the synthesis of **6.8** is described. Going back to the original work [*J. Am. Chem. Soc.*, **106**, 7352-7359 (1984)] he was able to find the melting point of compound **6.8**, i.e. 124–125 °C.

The same search could be carried out online (or in a computerized way). One has to remember that chemical compounds can be searched online only by using their CAS RN. Thus our student would have first to identify the CAS RN of 2(5*H*)-furanone, 3,4,5-triphenyl-. This could be done using the Registry File of CAS ONLINE (via STN) or using the Dictionary Chemical File of other vendors (e.g. Chemname of Dialog, Chemline at DIMDI, Chemdex at Orbit, Canon at Telesystem Questel). This particular search on STN is shown below:

⇐ FILE REG
FILE 'REGISTRY' ENTERED AT 01:37:02 ON 18 JUL 86
COPYRIGHT 1986 BY THE AMERICAN CHEMICAL SOCIETY

⇐ S "2(5H)-FURANONE, 3,4,5-TRIPHENYL-"/CN ⟵——————(1)
L1   1 "2(5H)-FURANONE, 3,4,5-TRIPHENYL-"/CN

⇐ D L1 SUB CAN ⟵————————————————————(2)
L1 ANSWER 1 OF 1

RN 7404-46-8 ⟵——————————————————————(3)
IN 2(5H)-Furanone, 3,4,5-triphenyl- (8CI, 9CI) ⟵—————(4)
SY 3,4,5-Triphenylfuran-2(5H)-one ⟵————————(5)
MF C22 H16 02 ⟵————————————————————(6)
     *
     * ⟵—————————————————————————(7)
     *
     *

10 REFERENCES IN FILE CA (1967 TO DATE) ⟵————————(8)
1   CA104(17):148100u
2   CA101(25):229794p
3   CA96(5):34748x
4   CA90(3):21939m
5   CA87(11):84168t
6   CA83(21):178668a
7   CA81(17):105127x
8   CA78(13):84167f
9   CA74(21):111828y
10   CA74(11):53359h

⇐ FILE CA
FILE 'CA' ENTERED AT 01:38:06 ON 18 JUL 86
COPYRIGHT 1986 BY THE AMERICAN CHEMICAL SOCIETY

⇐ S L1/P
L2 6 L1/P ⟵————————————————————————(9)

⇐ D L2 ⟵——————————————————————————(10)

L2   ANSWER 1 OF 6

AN   CA101(25):229794p
TI   Photochemical and thermal transformations of 2-(3H)-furanones and
     bis(benzofuranones). A laser flash photolysis study
AU   Lohray, B. B.; Kumar, C. V.; Das, P. K.; George, M. V.
CS   Dep. Chem., Indian Inst. Technol.
LO   Kanpur 208016, India
SO   J. Am. Chem. Soc., 106(24), 7352-9
SC   22-8 (Physical Organic Chemistry)
DT   J
CO   JACSAT
IS   0002-7863
PY   1984
LA   Eng

⇐ LOG Y
STN INTERNATIONAL LOGOFF AT 01:38:39 ON 18 JUL 86

(1) The Search command is used to search for the substance '2(5*H*)-furanone,
     3,4,5-triphenyl-' in the /CN (Chemical Name) search field. Quotation marks
     are used in order to mask the ( ) symbols in the name (other symbols which
     need masking are /> < =?!$'" as well as the Boolean operators AND OR
     NOT). The system finds one compound with that Chemical Name and places
     it in answer set L1.
(2) The Display command is used to display the compound in the SUBstance
     format followed by the latest ten (or less) numbers of the abstracts of the
     articles in which the compound was reported (CAN).
(3) The CAS RN is displayed in the RN field.
(4) The *CA* Index Name is displayed in the IN field.
(5) The synonyms for the substance are displayed in the SY field (up to 50 syn-
     onyms are displayed).
(6) The molecular formula for the substance is displayed in the MF field.
(7) The structure diagram for the substance is displayed (not shown here). The
     display may appear in a text-type (using ASCII terminal) or in a graphic (using
     Textronic or Textronic emulating terminal) form.
(8) The system informs us that the compound was referred to ten times in the *CA*
     File since 1967. The requested abstracts numbers are given.
(9) The search is continued in the *CA* File. It could be carried out by using the
     CAS RN of our compound followed by the suffix P (S 7404-46-8P), indicating
     that we are searching for the preparation of the compound. Another method
     is to use the answer set (L1) from the REG File and '/P' to indicate that we
     want only preparative information. Six syntheses of the compound have been
     reported since 1967; all are placed in answer set L2.
(10) We ask to Display the first answer in a bibliographic formate (the system
     default).

     The first answer is Lohray's work which was found manually; it reports a
m.p. of 124–125 °C for compound **6.8**.

6.3.2. Inorganic and organometallic compounds

How can the melting points of inorganic and organometallic compounds be located? Let us suppose that we are interested in finding out the melting points of magnesium nitride (**6.28**), xenon difluoride (**6.29**), and ferrocene-alpha-methylpropionic acid (**6.30**). The first place to check would be the *CRC Handbook of Chemistry and Physics*. Here one can find two large tables which may contain the desired information: Physical Constants of Inorganic Compounds and Physical Constants of Organometallic Compounds. On examining the first table we can locate the desired information concerning compound **6.28**. Under Magnesium there is data related to magnesium nitride (compound m 64). The decomposition point of the compound is reported, i.e. 800 °C. Compound **6.29** is not mentioned in this table and compound **6.30** is not reported in the second table.

The fact that the *CRC Handbook* (the same is true of any other desk handbook) contains data about only a small number of compounds should not surprise us. Handbooks are supposed to provide a rapid and convenient entry to various kinds of data, while their price should be reasonable in order to enable the individual chemist to purchase them for his personal use. All this limits the size and coverage of a handbook. One should not expect to find in a handbook any information dealing with 'exotic' compounds or even information related to a large number of 'non-exotic' compounds.

Mg_3N_2 XeF_2

6.28   **6.29**

6.30

Where could we find the desired information on compounds **6.29** and **6.30**? Is there any source that can be used before searching the *CA*? One such source for organometallic compounds is the *Dictionary of Organometallic Compounds* which was mentioned earlier (p. 85). Unfortunately the library did not hold the *Dictionary of Organometallic Compounds*. It was decided before turning to another library which holds this *Dictionary* to see if there is another source in the library that can be used. Indeed, there is another source; it is *Gmelin's Handbuch der Anorganischen Chemie* which covers the areas of inorganic and organometallic chemistry in a similar way to the coverage of organic chemistry in the *Beilstein*.

6.3.2.1. Gmelin's Handbuch der Anorganischen Chemie

Gmelin's Handbuch der Anorganischen Chemie (*Gmelin's Handbook of In-*

organic Chemistry) is the most comprehensive and the most modern collection of data on all inorganic substances. The information reported in the compendium is of value not only to inorganic and physical chemists but also to geochemists, physicists, metallurgists, technologists, mineralogists, and crystallographers.

The first edition of the *Handbook* was published by Leopold Gmelin in 1817, entitled *Handbuch der Theoretischen Chemie* (*Handbook of Theoretical Chemistry*). The work was composed of of three volumes, totalling 1588 pages of text. The fourth edition, which was published in the years 1848–66, was translated into English immediately after its publication by the Cavendish Society. Until its fifth edition the *Handbook* covered the total knowledge of chemistry at the time of publication. In the fifth edition organic chemistry was removed from the *Handbook* and since then it has been covered in a different compendium—*Beilstein's Handbuch der Organischen Chemie*.

The last edition of the *Gmelin* to be published is the eighth edition, whose publication started in 1922. In this edition one volume (in most cases in several parts) is devoted to each element. Towards the end of the publication of the eighth edition (the end of the 1950s) it was decided not to proceed with the publication of a new edition, but instead to turn the *Handbook* into an 'open-ended' publication, with publication of Supplementary Volumes related directly to specific volumes of those elements whose coverage was completed a long time ago. Each of the volumes published since then has a literature coverage of up to a few months prior to its date of publication (e.g. volumes published in 1986 have a complete coverage of the literature at least up to the end of 1984 and usually include some of the important publications of 1985). At the beginning of the 1970s the publication of Volumes of the New Supplement to the eighth edition began. These are volumes dealing with important modern topics in inorganic chemistry, not necessarily related to a specific element (e.g. Water Desalting, Rare Gases Compounds, Carboranes).

By the end of 1986 about 520 separate volumes (not including the index volumes) of the *Handbook* had been published (an average of 20 volumes were published yearly during the last ten years—1976–1986. Up to the end of 1977 the *Handbook* was divided into three parts:

(a) 71 Main Volumes arranged according to the Gmelin System Number (numbered from 1—rare gases to 71—transuranium elements).
(b) Supplement Volumes related to specific Main Volumes.
(c) Volumes of the New Supplement to the eighth edition.

At the beginning of 1978 a new organization of the *Handbook* took place. The individual volumes (Main Volumes, Supplement Volumes, and Volumes of the New Supplement) were arranged alphabetically by chemical symbols from Ac (actinium) to Zr (zirconium). Thus 'Rare Gases Compounds' (Volume 1 of the New Supplement) is now placed under He (rare gases—He,

Ne, Ar, Kr, Xe, Rn—first element He). Each new volume is related to its respective element.

Gmelin gives critically evaluated data about the elements and their compounds, including many tables of numerical data and graphic material (graphs, diagrams of various apparatus, etc.). All of the information is followed by references to the original literature.

A System Number is assigned to each element (except the rare gases, rare earths, and transuranium elements—to each one of these three groups one System Number has been assigned) and to ammonia. Anion-forming elements have been given smaller numbers than cation-forming elements; the Gmelin System is described in Table 6.5. The various inorganic compounds are arranged in *Gmelin* according to the Principle of Latest Possible Entry. Thus a compound or a combination of elements is discussed in the volume assigned to the element having the highest System Number (as in *Beilstein*). For example, HCl is discussed in the volume assigned to chlorine (the System Number of H is 2 and that of Cl is 6), zinc chloride is discussed in the volume devoted to zinc (the System Number of Zn is 32), and both chromic chloride and zinc chromate are discussed in the chromium volume (the System Number of Cr is 52).

There is a Cumulative Formula Index in 12 volumes covering the Main Work and its Supplements up to the end of 1974 as well as Volumes 1–12 of the New Supplements (up to the end of 1973). There is a First Supplement to the Formula Index covering the period 1973/4–79. The Formula Index is arranged in an alphabetical order in which carbon and hydrogen are not treated separately (e.g. dibromochloromethane appears under Br_2CClH). An online cumulative index to the *Gmelin* became available at the beginning of 1987 on STN International (as File *GFI*). The File contains besides the Cumulative Formula Index also the Complete Gmelin Catalog. At this stage the coverage period of the online version of the index is identical to that of the printed one (the end of 1979). Plans are to update the file every six months.

The primary language used in the *Gmelin* was German. Since 1957 there has been a gradual changeover to English—English tables of contents, marginal keywords in English, and short English summaries of some sections were included in the various volumes. Reprints with the English additions to volumes published prior to 1957 are available. (When you are using the *Gmelin* note the date of the original publication of the volume and the latest date of the literature coverage. These dates are to be found on the reverse side of the title page. Do not mix it up with the 'reprint' date. The textual material of the 'reprint' is identical with that of the original publication.) During the late 1970s and early 1980s various chapters and even whole volumes (e.g. Water Desalting, Radium Supplement 2) were published in English. At the beginning of 1982 there was a complete changeover into English and any volume published from 1982 onwards was published in English.

Table 6.5 *Gmelin* system numbers

System No.	Element	System No.	Element
1	Rare Gases	37	Indium
2	Hydrogen	38	Thallium
3	Oxygen	39	Rare earths
4	Nitrogen	40	Actinium
5	Fluorine	41	Titanium
6	Chlorine	42	Zirconium
7	Bromine	43	Hafnium
8	Iodine	44	Thorium
9	Sulphur	45	Germanium
10	Selenium	46	Tin
11	Tellurium	47	Lead
12	Polonium	48	Vanadium
13	Boron	49	Niobium
14	Carbon	50	Tantalum
15	Silicon	51	Protactinium
16	Phosphorus	52	Chromium
17	Arsenic	53	Molybdenum
18	Antimony	54	Tungsten
19	Bismuth	55	Uranium
20	Lithium	56	Manganese
21	Sodium	57	Nickel
22	Potassium	58	Cobalt
23	Ammonium	59	Iron
24	Rubidium	60	Copper
25	Cesium	61	Silver
26	Beryllium	62	Gold
27	Magnesium	63	Ruthenium
28	Calcium	64	Rhodium
29	Strontium	65	Palladium
30	Barium	66	Osmium
31	Radium	67	Iridium
32	Zinc	68	Platinum
33	Cadmium	69	Technetium
34	Mercury	70	Rhenium
35	Aluminium	71	Transuranium elements
36	Gallium		

How can we locate our desired information concerning XeF_2 using the *Gmelin*? The first step is to consult the Cumulative Formula Index. Looking under F_2Xe (alphabetical order) we found—preparation Erg. W1 26/9 Erg. means New Supplement Series, and W1 means Volume 1. Thus all the information concerning the preparation of XeF_2 is reported in Volume 1 of the New Supplements (Rare Gases Compounds) on pages 26–29. Turning to the entry, we find a critical review of the preparation of XeF_2 with many references, but it does not report the desired melting point. However, let us choose two of the references given in the above review for further search.

One of them is Maln's well known review of rare gas compounds [*Chem. Rev.*, **65**, 199 (1965)], the other is Falconer's work on the photochemical synthesis of XeF_2 (*J. Inorg. Nucl. Chem.*, **29**, 1380 (1967)). Table 2 of the first reference gives physical properties of XeF_2, including its m.p.-140 °C—while the second reference gives a m.p. of 134 °C.

One can also use *Gmelin* to retrieve the m.p. of compound **6.30**. The first step would be to examine the first supplement to the Formula Index. Under the formula $C_{14}FeH_{16}O_2$ there are 12 entries, each one with a semistructural formula. The sixth entry is $C_5H_5FeC_5H_4CH_2CH(CH_3)COOH-Fe$ Org. Verb. A3$-71/2$, 75, 100/1. Org. Verb. stands for Organische Verbindungen (Organic Compounds); Organoiron Compounds A3 (Fe Organische Verbindungen) is Volume 50 of the New Supplements; there on page 72 compound **6.30** appears as compound 4 in Table 10 and its melting point is reported to be 110–12 °C.

6.4.   Locating crystallographic data

6.4.1. Space groups and lattice constants

A crystallographer is interested in the space group and lattice constants of biphenyl (**6.31**) and 4,4'-dibromobiphenyl (**6.32**).

6.31

6.32

Where should he look for the desired data? The first place one should check is the *CODATA Directory of Data Sources for Science and Technology*.

6.4.1.1. CODATA Directory of Data Sources for Science and Technology

The publication of the *CODATA Directory of Data Sources for Science and Technology* started in 1977; it is a compilation of international and national data projects, data centres, computerized data bases, publication series, and other bibliographic entries to numerical data sources. The *CODATA Directory* serves as an information source on where to find numerical data such as properties and behaviour of matter, numerical values of chemical and biological systems, various experimental numerical values, etc. It is published chapter by chapter in the *CODATA Bulletin*, up to the end of 1985 twelve chapters have been published (Table 6.6). Chapters 1, 6, 8, 11, and to some extent Chapters 2, 9, and 12 are of interest to chemists.

Our crystallographer should use the Crystallography Chapter of the *CODATA Directory* (this chapter was published in June 1977 as a *CODATA Bulletin*, No. 24) as a general information source.

Examining the Crystallography Chapter, our crystallographer, who is interested in the crystal structure of organic compounds, found a few sources that should be used: *Landolt-Bornstein's Zahlenwerte und Funktionen aus Physik, Chemie, Astronomie, Geophysik, und Technik*; *Landolt–Bornstein's Zahlenwerte und Funktionen aus Naturwissenschaften und Technik, Neue Serie; Molecular Structures and Dimensions*, edited by O. Kennard, which is the printed output of the Cambridge Crystallographic Data Centre; and *Crystal Data Determinative Tables*, edited by J.D.H. Donnay and H.M. Ondik, which is the printed output of the NBS Crystal Data Center.

6.4.1.2. Landolt-Bornstein's Zahlenwerte und Funktionen aus Physik, Chemie, Astronomie, Geophysik und Technick; Landolt-Bornstein's Zahlenwerte und Funktionen aus Naturwissenschaften und Technik, Neue Serie

Landolt-Bornstein's Zahlenwerte und Funktionen aus Physik, Chemie, Astronomie, Geophysik und Technik (*Landolt–Bornstein's Numerical Data and Functional Relationship in Physics, Chemistry, Astronomy, Geophysics and Technology*) is the most comprehensive printed compilation of numerical data gathered from the various fields of physical and technological sciences. This compendium, better known simply as *Landolt-Bornstein*, started in 1883 as the *Physikalisch-Chemischen Tabellen*, which was compiled by H.H. Landolt and R. Bornstein and totalled 261 pages. The sixth and last edition was published during the years 1950–80. It is composed of 28 parts in four volumes (Atomic and Molecular Physics, Properties of Matter in Aggregated State, Astronomy and Geophysics, and Technology). The first two volumes are of interest to the chemist. In 1961 the first part of the New Series (*Landolt-Bornstein's Zahlenwerte und Funktionen aus Naturwissenschaften und Technik, Neue Serie*) was published. The New Series is planned to be a flexible 'open-end' publication. Up to the end of 1986 over 110 volumes and subvolumes arranged in six subject groups (Nuclear and Particle Physics, Atomic and Molecular Physics, Crystallography and Solid State Physics, Macroscopic and Technical Properties, Geophysics and Space Research, and Astronomy and Astrophysics) have been published. It updates some of the topics and subjects that have been covered by the sixth edition. However, it mainly covers currently expanding fields of research which were not covered before. Because of the flexibility of the publication supplementary volumes of the whole or specific aspects of a topic can be brought out as needed (e.g. Volumes II/9, II/10 of the sixth edition are devoted to magnetic properties, Volume III/4 of the New Series deals with magnetic and other properties of oxides and related compounds. This subject was updated and extended with the publication of Volumes III/12/a, III/12/b, and III/12/c in the years 1978–80, entitled Garnets and Perovskites; Spinels, Fe Oxide, and Fe-Me-O Compounds; and Hexagonal Ferrites, and Special Actinide and Lanthanide Compounds, respectively). Three of the six groups of the new series are of interest to chemists—Group II, Group III and Group IV (Atomic and Molecular Physics, Crystallography and Solid State Physics, and Macroscopic and Technical Properties, respectively). The language used in the sixth edition is German (except Volumes II/9 and II/10 which are bilingual—German–English). In contrast English is used extensively in the New Series, either as the only language (in most cases) or side by side with German. Indexes to the sixth edition as well as to the New Series are in preparation.

How can our crystallographer obtain the data he is after using the *Landolt-Bornstein*?

The first stage is to find out where in the *Landolt–Bornstein* the information is located. This can be found by examining Group III of the New Series which is devoted to Crystallography and Solid State Physics. Examining the various volumes and subvolumes of Group III one finds Volume II/5/a entitled Structure Data of Organic Crystals C_1-C_{13}, and Volume III/5/b entitled Structure Data of Organic Crystals C_{14}-C_{140}. The empirical formula of compound **6.31** is $C_{12}H_{10}$ and that of compound **6.32** is $C_{12}H_8Br_2$. Thus the desired information should be searched for in Volume III/5/a. Our crystallographer indeed found compound **6.32** on page 631 reported as compound 12-8-7, where he obtained the following information: $C_{12}H_8Br_2$, 4,4'-dibromobiphenyl (312.0) C^5_{2h}, $P2_{1/n}$, $a = 15.73$, $b = 14.8$, $c = 9.83$, $\beta = 96.07°$, $z = 8$. He also found two references which he kept for his records [J. Dhar, *Ind. J. Phys.*, **20**, 154 (1946); F.H. Herbstein, *Acta Crystallogr.*, **11**, 666 (1958)]. Compound **6.31** was found on page 645 reported as compound 12-10-2: $C_{12}H_{10}$ biphenyl (154.2) C^5_{2h}, $P2_{1/a}$ $a = 8.12 \pm 2$, $b = 5.63 \pm 1$, $c = 9.51 \pm 2$, $\beta = 95.1 \pm 0.3°$, $z = 2$. The data were accompanied by four references and lattice constants which were measured at -180 °C; all this information was recorded by our crystallographer.

6.4.1.3. Molecular Structures and Dimensions

Molecular Structure and Dimensions is a classified bibligoraphy of crystal structures of organometallic and organic compounds prepared by the Crystallographic Data Centre at Cambridge University. The printed bibliography is published by Oosthoen, Utrecht, in conjuction with the International Union of Crystallography. Volumes 1 and 2 cover the period 1935–69. Subsequent volumes cover approximately a one year period each. At the end of each volume there is a set of indexes (Compound Name, Formula, Permutated Formula, Author, and Literature). There is a cumulative index in two volumes for Volumes 1–8, covering the period 1935–75.

Our crystallographer decided to use first the Cumulative Formula Index of Volumes 1–8. Here he found that only formulae are given, without compound names, thus not enabling him to decide if any entry reported is an entry to the desired compound (biphenyl in the case of $C_{12}H_{10}$) or to any other compound with the same empirical formula (in the case of $C_{12}H_{10}$ it may be 1,2-dihydroacenaphtalene, 1,2-dihydro-1,2-bismethylenenazulene, 1,2-dihydrocyclobuta[*b*]naphthalene, benzocyclooctatetraene, heptalen, etc). He decided to use the Compound Name Index, which is a KWIC index that enables him to locate biphenyl and its dibromo derivative simultaneously. Using the Cumulative Compound Name Index he could not find any entry to compound **6.32** but there were four entries to compound **6.31**. Out of those entries two were reported in *Landolt–Bornstein*, the third was from 1976, describing measurements taken at 110 K, and the fourth was an entry from 1961 [G.B. Robertson, *Nature (London)*, **191**, 593 (1961)], that for some reason was not reported in *Landolt–Bornstein*. As there are no cumulative indexes for Volume 9 onwards, our crystallographer had to examine each index volume separately. Let us see what he found, for example, in Volume 9. There was one entry to compound **6.31** [G.P. Charbonneau and Y. Delugeard, *Acta Crystallogr.*, Section B, **33**, 1586 (1977)], and one entry to compound **6.32** [P. Kronebusch, W.B. Gleason and D. Britton, *Cryst. Struct. Commun.*, **5**, 839 (1976)].

Molecular Structures and Dimensions is also available in machine readable form either via various vendors (e.g. CIS, CISTI) or for inhouse usage. An inhouse search of the Cambridge tapes for compound **6.32** gave one reference as can be seen below.

INPUT...Q *FORMUL 'C12 H8 BR2' AND *COMPND 'BIPHENYL'

NUMBER OF QUESTIONS 1
NUMBER OF STRINGS 2
TOTAL NUMBER OF CHARACTERS 20

4,4'-DIBROMO-BIPHENYL
C12 H8 BR2

P.KRONEBUSCH,W.B.GLEASON,D.BRITTON, CRYST.STRUCT.
COMMUN., 5,839,1976. 189
DBRBIP   770506 9.19.8
1/19
* END OF FILE REACHED ON UNIT 1 *

NUMBER OF ENTRIES SEARCHED = 39497

The full numerical data could also be retrieved:

CELL	9.760 14.110 17.430		90.000 116.600	90.000			
SYMM	1.	0. 0.	0.0	0.	1.	0.	0.0	0. 0. 1. 0.0
SYMM	−1.	0. 0.	0.0	0.	1.	0.	0.50000	0. 0. −1. 0.50000
SYMM	−1.	0. 0.	0.0	0.	−1.	0.	0.0	0. 0. −1. 0.0
SYMM	1.	0. 0.	0.0	0.	−1.	0.	−.50000	0. 0. 1. − .50000

BR1	0.01690	−0.16720	0.10900
C2	0.01990	−0.03310	0.11330
C3	0.15640	0.01040	0.16430
C4	0.15430	0.11210	0.16950
C5	0.02230	0.16190	0.12370
C6	−0.11100	0.11560	0.07070
C7	−0.11360	0.01430	0.06580
C8	0.02140	0.26760	0.12990
C9	0.10060	0.31140	0.20710
C10	0.09800	0.40990	0.21620
C11	0.01160	0.46110	0.14330
C12	−0.07210	0.42110	0.06480
C13	−0.06790	0.32240	0.05820
BR2	0.00250	0.59570	0.15500
BR1'	−0.48000	0.08360	0.10120
C2'	−0.49110	0.21610	0.11160
C3'	−0.35680	0.26700	0.13660
C4'	−0.36290	0.36730	0.14250
C5'	−0.50300	0.41190	0.12210
C6'	−0.63270	0.35740	0.09930
C7'	−0.62870	0.25930	0.09220
C8'	−0.50570	0.51460	0.12770
C9'	−0.42570	0.57350	0.09740
C10'	−0.43230	0.67100	0.10150
C11'	−0.52000	0.71160	0.13550
C12'	−0.60030	0.65800	0.16710
C13'	−0.59640	0.55800	0.16040
BR2'	−0.52600	0.84570	0.14290

ENTNO= 1 DBRBIP SPG=P21/C Z= 8 ATOMS= 28 INTF=3
AS=3 R−FACTOR(MIN,MAX)

BOND LENGTHS

BR1	C2	1.893	C2	C3	1.371	C2	C7	1.365	C3	C4	1.438
C5	C8	1.495	C6	C7	1.431	C8	C9	1.363	C8	C13	1.394
C11	C12	1.362	C11	BR2	1.916	C12	C13	1.399	BR1'	C2'	1.886

C3'	C4'	1.422	C4'	C5'	1.398	C5'	C6'	1.378	C5'	C8'	1.453
C8'	C13'	1.391	C9'	C10'	1.380	C10'	C11'	1.366	C11'	C12'	1.369

BOND ANGLES

BR1	C2	C3	117.8	BR1	C2	C7	118.2	C3	C2	C7	124.1 C
C4	C5	C6	120.5	C4	C5	C8	120.3	C6	C5	C8	119.2 C
C5	C8	C9	120.3	C5	C8	C13	120.7	C9	C8	C13	118.9 C
C10	C11	C12	123.7	C10	C11	BR2	117.5	C12	C11	BR2	118.8 C
BR1'	C2'	C3'	117.3	BR1'	C2'	C7'	120.6	C3'	C2'	C7'	122.1 C
C4'	C5'	C6'	119.3	C4'	C5'	C8'	118.5	C6'	C5'	C8'	122.2 C
C5'	C8'	C9'	122.6	C5'	C8'	C13'	120.0	C9'	C8'	C13'	117.3 C
C10'	C11'	C12'	121.7	C10'	C11'	BR2'	119.4	C12'	C11'	BR2'	118.2 C

TORSION ANGLES

BR1	C2	C3	C4	−177.3	C7	C2	C3	C4	2.5	BR1	C2	C7
C2	C3	C4	C5	−1.5	C3	C4	C5	C6	−0.9	C3	C4	C5
C8	C5	C6	C7	−178.2	C4	C5	C8	C9	−40.4	C4	C5	C8
C6	C5	C8	C13	−35.7	C5	C6	C7	C2	−1.6	C5	C8	C9
C5	C8	C13	C12	178.1	C9	C8	C13	C12	2.1	C8	C9	C10
C9	C10	C11	BR2	178.5	C10	C11	C12	C13	−0.8	BR2	C11	C12
BR1'	C2'	C3'	C4'	178.2	C7'	C2'	C3'	C4'	0.3	BR1'	C2'	C7'

C2'	C3'	C4'	C5'	−0.8	C3'	C4'	C5'	C6'	2.2	C3'	C4'	C5'
C8'	C5'	C6'	C7'	178.4	C4'	C5'	C8'	C9'	43.1	C4'	C5'	C8'
C6'	C5'	C8'	C13'	38.4	C5'	C6'	C7'	C2'	2.5	C5'	C8'	C9'
C5'	C8'	C13'	C12'	179.6	C9'	C8'	C13'	C12'	−3.5	C8'	C9'	C10
C9'	C10'	C11'	BR2'	178.9	C10'	C11'	C12'	C13'	−2.8	BR2'	C11'	C12

END OF FILE REACHED ON UNIT 1 ...   1 ENTRIES

When our crystallographer compared the results obtained from the two
sources it was clear that the coverage of *Landolt-Bornstein* is more complete than that of the *Molecular Structures and Dimensions*. On the other
hand, the period covered by the *Landolt-Bornstein* extended only to 1968,
whereas *Molecular Structures and Dimensions* includes data up to 1986 in its
printed form and is nearly completely up to date in its online version. Another difference between the two sources is that whereas *Landolt-Bornstein*

supplies the user with numerical data followed by bibliographic information (and sometimes critical remarks) on the values reported, *Molecular Structures and Dimensions* gives only bibliographic information, and in order to obtain numerical data one has to consult the original work; however all the crystallographical data are available on its machine readable version.

6.4.2. Powder diffraction data

If a mineralogist needs the X-ray powder diffraction spectra of pure zircon ($ZrSiO_4$), how would he be able to obtain the information? Examining the Crystallography Chapter of the *CODATA Directory* (see p. 105) he found two sources: Dana's *System of Mineralogy* and *Selected Powder Diffraction Data for Minerals*, which is a printed output of the Powder Diffraction File. He decided to use the second source.

The Powder Diffraction File contains powder diffraction data for organic, organometallic, metal, alloy, mineral and other inorganic compounds. The output of the file is available as data cards, microfiche, and as a data base on a magnetic tape. Various search manuals are available in order to enable the user to locate the desired data (e.g. *Search Manual—Hanawatt Method, Fink Method*, both for inorganic compounds; *Organic Search Manual*; *Selected Powder Diffraction Data for Minerals, Search Manual*). Using the Name Index our mineralogist would be able to find under zircon its formula, $ZrSiO_4$, the three strongest lines of powder pattern with their relative intensities 3.30_x, 4.43_5, and 2.52_5, the number of the corresponding data card, 6-266, and the number and coordinate of the microfiche card, 1-2-D12. Knowing the number of the data card and/or microfiche card the mineralogist should be able to obtain all the available powder diffraction data on zircon.

6.5. Spectral data

When a chemist is interested in some spectroscopic data of an organic or inorganic compound, he should first check the *Beilstein* or the *Gmelin*, respectively (however he should be aware of the time cutoff of the specific *Beilstein* Supplement or *Gmelin* Volume he is using). If the desired data are unavailable in the above sources, or are over 10–15 years old, one should turn to one of the various collections of spectra and/or to the *CA* itself.

Let us consider an inorganic chemist who needs the IR spectra of $H_4[Os(CN)_6]$ and who decided to use the *Gmelin* as the first source in his search. There are two volumes in *Gmelin* dealing with osmium and its compounds; Volume 66 of the eighth edition published in 1939 and having a literature closing date of August 1938 (a reprint of the volume was published in 1978), and Supplement Volume 1—Elements and Compounds, published in 1980 and having a literature closing date of the end of 1978, with more recent data included.

112

Our chemist decides to use the Supplement Volume first. He knew that the material in that volume was not indexed at the First Supplement to the Formula Index as it covers only volumes issued no later than the end of 1979. He decided therefore to examine the table of contents; in doing so he found that Part 3 deals with Osmium compounds, Chapter 3.11 is devoted to osmium-carbon compounds, and osmium cyanides are discussed in Section 3.11.2. This section is divided into two subsections, of which the first one is of interest—Unsubstituted Cyanides Complexes, page 173. Turning to page 173 our chemist found it to be devoted to $H_4[Os(CN)_6]$ (usually there is no volume index(es) to the various *Gmelin* volumes), where he found the desired information followed by two references—D.F. Evans, D. Jones, and G. Wilkinson, *J. Chem. Soc.*, 3164 (1964) and A.P. Ginsberg and E. Koubek, *Inorg. Chem.*, **4**, 1186 (1965). Our chemist tried to get newer data concerning the compound only to find that since Ginsberg's work no new data on $H_4[Os(CN)_6]$ has been reported in the literature.

On many occasions, using *Beilstein* or *Gmelin* saves a lot of search time and gives the desired information instantly. Such was the case with an organometallic chemist who urgently needed the IR spectra of the coordination compound of pyridine, 2-methyl-4-nitro-1-oxide (**6.33**) with $La(ClO_4)_3$.

6.33

The chemist decided to use the *CA* for his search. However he had a problem, how would he conduct his search? He did not know the stoichiometry of the coordination compound and therefore decided not to use the *CA* Formula Index. Using the *CA* Chemical Substance Index he could not find any entry to lanthanum perchlorate alone (not to mention its various coordination compounds). This looks very strange to our chemist as he was sure that the lanthanum perchlorate is a well known compound (he knew that it is commercially available). He decided to turn to Appendix IV of the Index Guide, an appendix entitled Chemical Substance Index Name. There he found rules about the *CA* nomenclature of various salts and coordination compounds.

Salts of inorganic cations with organic or inorganic acids are indexed in the *CA* Chemical Substance Index under the acid, e.g. $Ca(CH_3COO)_2$—acetic acid calcium salt, $Ca(MnO_4)_2$—permanganic acid calcium salt. In the *CA* Formula Index the following entries will be used respectively—$C_2H_4O_2$ calcium salt, $HMnO_4$ calcium salt. Salts of organic cations with organic or inorganic acids are indexed in the *CA* Chemical Substance Index under the cation name, e.g. $(C_4H_9)_4NMnO_4$—butanaminium, N,N,N-tributyl salt with permanganic acid. In the *CA* Formula Index it will appear as $C_{16}H_{36}N$, salt with permanganic acid.

Another thing our chemist found in the above Appendix was that molecular addition compounds of neutral components are indexed in the *CA* Chemical Substance Index and in the *CA* Formula Index under the name and the formula of each of the components. Knowing this he decided to search again in the *CA* Chemical Substance Index. He realized that the coordination compound he is looking for will appear under perchloric acid, lanthanum compound (or complex) as well as under pyridine, 2-methyl-4-nitro-1-oxide, lanthanum compound (or complex). He decided to use the first entry, searching the 1985–82 (Volumes 96–103) *CA* Chemical Substance Index as well as the 10th Collective Chemical Substance Index. Using the single volume indexes our chemist found 6 entries with perchloric acid, lanthanum complex, each one with a different RN. In order to identify the various compounds our chemist turned to the *CAS Registry Handbook— Number Section*. This part of the Handbook connects the CAS RN and the compounds *CA* Index Names and molecular formulae. None of the six different compounds were the one he was looking for. In the 10th Collective Chemical Substance Index, nine entries were found under perchloric acid, lanthanum complex; one of them having a CAS RN 74345-07-6 turned out to be the desired compound—lanthanum (+2) pentakis (2-methyl-4-nitropyridine-1-oxide-01) (perchlorato-O) ($C_{30}H_{30}ClLaN_{10}O_{19}.2ClO_4$). This compound is reported in *CA* 93:60135c. It is a work of N.S. Navaneetham and S. Soundararajan in *Proc. Indian Acad. Sci. (Ser) Chem. Sci.*, **89**, 17–23 (1980). This journal is not a common one and efforts should be made to get it as it contains the data needed (as it is written in the abstract— '... were prepd. and characterized by anal., electrolytic conductance, and IR, ^{1}H NMR and...'). The search time could have been shortened if our chemist was using the other chemical name of the compound—pyridine, 2-methyl-4-nitro-1-oxide, lanthanum complex. This kind of compound is not reported at all during the years 1985–82, and reported only once during the 10th Collective Index (CI) period (CAS RN 74345-07-6; CA 93:60135c). The search could be shortened even more by running an online search.

FILE 'REGISTRY' ENTERED AT 01:50:24 ON 18 JUL 86
COPYRIGHT 1986 BY THE AMERICAN CHEMICAL SOCIETY

⇐ S PERCHLORIC ACID, LANTHANUM COMPLEX/CN ⟵————(1)
L1   39 PERCHLORIC ACID, LANTHANUM COMPLEX/CN

⇐ S L1 AND PYRIDINE, 2-METHYL-4-NITRO-, 1-OXIDE, LANTHANUM COMPLEX/CN
          2 PYRIDINE, 2-METHYL-4-NITRO-,1-OXIDE,
          2 LANTHANIM COMPLEX/CN
L2   2 L1 AND PYRIDINE, 2-METHYL-4-NITRO-,
          2 1-OXIDE, LANTHANUM COMP

⇐ D L2 1-2
L2   ANSWER 1 OF 2

RN   74345-07-6
IN   Lanthanum(+ 2), Pentakis(2-methyl-4-nitropyridine
     1-oxide-01)(perchlorato-O)-, (OC-6-22)-, diperchlorate (9CI)
SY   Pyridine, 2-Methyl-4-nitro-, 1-oxide, lanthanum complex (9CI)
SY   Perchloric acid, lanthanum complex (9CI)
MF   C30 H30 Cl La N10 O19 . 2 Cl O4
        *
        *
        *
        *

1 REFERENCES IN FILE CA (1967 TO DATE)

L2   ANSWER 2 OF 2

RN   74345-06-5
IN   Lanthanum(2 +), pentakis(2-methyl-4-nitropyridine
     1-oxide-01)(perchlorato-O)-, (OC-6-22)- (9CI)
SY   Pyridine, 2-methyl-4-nitro-, 1-oxide, lanthanum complex (9CI)
SY   Perchloric acid, lanthanum complex (9CI)
MF   C30 H30 Cl La N10 O19
CI   CCS, COM   ⟵───(2)
ST   7:0C-6-22 ⟵───(3)
        *
        *
        *
        *

0 REFERENCES IN FILE CA (1967 TO DATE)   ⟵───────────────(4)

⇐ FILE CA ⟵──(5)
FILE 'CA' ENTERED AT 01:52:49 ON 18 JUL 86
COPYRIGHT 1986 BY THE AMERICAN CHEMICAL SOCIETY

⇐ S L2 ⟵───(6)
L3   1 L2

⇐ D L3 BIB ABS ⟵───(7)

L3   ANSWER 1 OF 1

AN   CA93(6):60135c
TI   4-Nitro-2-picoline 1-oxide complexes of lanthanide perchlorates

AU Navaneetham, N. S.; Soundararajan, S.
CS Dep. Inorg. Phys. Chem., Indian Inst. Sci.
LO Bangalore 560 012, India

SO Proc. - Indian Acad. Sci., [Ser.]: Chem. Sci., 89(1), 17-23
SC 78-7 (Inorganic Chemicals and Reactions)
DT J
CO PIAADM
PY 1980
LA Eng

AB 4-Nitro-2-picoline 1-oxide (NPicO) complexes of the formulas
    La(NPicO)5(ClO4)3, Ln2(NPicO)9(ClO4)6 Ln = Pr, Nd, Gd) and
    Ln (NPicO)4(ClO4)3 (Ln = Tb, Dy, Ho, Yb were prepd.
    and characterized by anal., electrolytic conductance,
    and IR, 1H NMR, and electronic spectra data. A tentative
    coordination no. of 6 for all of the complexes is
    assigned.

⇐ LOG Y
STN INTERNATIONAL LOGOFF AT 01:54:10 ON 18 JUL 86

(1) The Search command is used to search for the perchloric acid, lanthanum complex in the /CN search field. 39 complexes are found by the system and placed in answer set L1. The Search command is used to combine (using the Boolean operator AND) answer set L1 and pyridine, 2-methyl-1-nitro, 1-oxide, lanthanum complex in the /CN search field. There are two such complexes in the *REG* File.

(2) The Class Identifier (CI) field indicates that the compound is a coordination compound (CCS) and also occurs as a component of a mixture (COM).

(3) The Stereochemical code (ST) field includes stereochemical information on the substance which is not searchable.

(4) There are cases of chemical compounds reported in the REG File and not reported in the literature. One class of compounds are cations or anions of bases or acids. Here we have the cation of the compound having CAS RN 74345-07-6.

(5) After locating the substance information needed in the REG File, we change to the *CA* File to search for bibliographical data.

(6) The Search command is used to search for the substance(s) we found in the REG File in the *CA* File.

(7) We ask to display the only answer in answer set L3 in a BIBliographic and ABStract formate. The CAS does not license the usage of abstracts to vendors. However, they are searchable and displayable on STN International.

Whatever search route is used one remains with the problem of having to get a copy of the original work in order to record the required data.

It is really worth while to see if the desired information is reported in the *Gmelim* and could be retrieved from there. Volume D-2 of the Rare Earths is devoted to Coordination Compounds and was published in 1982; its literature coverage cutoff is the end of 1980. Our chemist may use the table of contents to look for complexes of pyridines, 2-methyl-4-nitro-1-oxide or he may use the Empirical Formula Index of ligands at the end of the volume (which is a formula index arranged according to the Hill System). In both cases one is referred to page 142. Indeed in page 142 one may find the IR spectra data of $La(ClO_4)_3.5C_6H_6N_2O_3$ as well as the reference to the Indian work.

One may see that using the *Gmelin* not only retrieves bibliographic information in a very short time (a few minutes), but also retrieves the desired numerical data—1295s, 841m, 800m, 1525s, 1135sh, 1045sh, 940s, 636sh, 110vs, 625m (s—strong, m—medium, sh—shoulder, vs—very strong).

In case one is searching for the ^{77}Se NMR of benzoselenophene (**6.34**) one should try first to get the information using the *Beilstein*. Compound **6.34** is reported in Volume 17; it is a chalcogen derivative of benzofuran. Searching for compound **6.34** in the Cumulative Formula Index of Volumes 17 and 18 one finds that the compound is reported in 17 EIII/IV 491. Using this as a coordination reference for the Fifth Supplement helped us to locate data on compound **6.34** in 17/2 EV 15; there one may find a reference to the ^{77}Se NMR of the compound [L. Christiaens *et al.*, *Org. Mag. Reson.*, **8**, 354 (1976)].

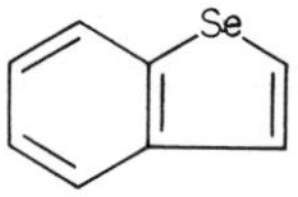

6.34

One may use SANDRA to locate compound **6.34** in the *Beilstein*. After drawing **6.34** using the software one enters the command Q (Quit—end of structure input). The program responds with "1 = correction; 2 = start analysis"; choosing 2 gave an immediate response "** please wait a few seconds**". In less than a minute the answer is displayed on the screen (Fig. 6.1).

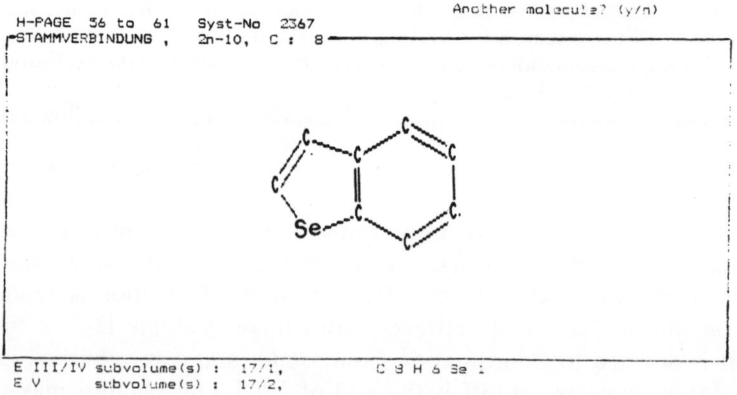

Figure 6.1   A SANDRA search of compound **6.34**

Upon examination of Volume EV 17/2 for a compound having $C:8$, a system number 2367 and unsaturation of 2n-10, one finds that the above compound if reported in the EV of *Beilstein* would be placed in Volume 17/2 p. 3–18. Indeed, looking through these fifteen pages results in locating the compound on page 15.

Some spectroscopic data could be located using the fifth edition of *Heilbron's Dictionary of Organic Compounds* (see p. 84). Let us consider a pharmaceutical chemist interested in the mass spectrum of chlorophentermine (4-chloro-alpha,alpha-dimethylbenzeneethaneamine) (**6.35**). As he was not able to find the desired data in the *Beilstein*, he decided to use *Heilbron*. He was able to locate the desired compound under Chlorphentermine (compound C-02335). He found there some information about the compound— its CA Index Name, the fact that the compound is used as an anorexic drug and appetite suppressant, its b.p., the m.p. of its hydrochloride, and five references. From the mnemonic suffixes to these references our chemist learned that one reference reports the desired data [A. Cartoni *et al.*, *Bull. Soc. Chim. Fr*, **115**, 547 (1976) (ms)]. Turning to the original work he was able to find the desired mass spectrum.

6.35

Other sources that should be checked when one is looking for spectroscopic data of organic compounds are the *CRC Atlas of Spectral Data and Physical Constants for Organic Compounds* edited by J.G. Grasselli and W.M. Ritchey and/or *CRC Handbook of Data on Organic Compounds* edited by R.C. Weast and M.J. Astle, both published by CRC Press. The first book provides in its latest edition data for about 21,000 compounds with entries to the original literature. The data included the usual physical constants (m.p., b.p., density, etc.) together with spectral data such as IR, Raman, UV, ^{1}H NMR, ^{13}C NMR, MS. Each entry has at least one spectral data element, the spectral data being listed in a tabular form. The second of these CRC products has been discussed earlier (see p. 82). Using those sources one has to keep in mind that the data were compiled mainly from secondary sources and were not critically evaluated.

6.5.1. ESR spectra

How could one find the ESR data (g factor and a values) of the anion radical derived from 1,2-diphenylacetylene (**6.36**)?

6.36

The first place to look should be the New Series of *Landolt–Bornstein*. Seven volumes and subvolumes—Volumes II/1, II/9a, II/9b, II/9c$_1$, II/9c$_2$, II/9d$_1$, and II/9d$_2$—report the magnetic properties of free radicals. The Substance Index to Volume II/1 and II/9a–II/9d$_2$ is located in Volume II/9d$_2$. Using this index one can find that compound **6.36** is reported in Volume II/9d$_1$, page 826, where the data related to its anion radical, data which was obtained by various chemical and electrochemical reduction methods at various temperatures (203–298 K), is given. The data includes references, *g* factors, and assignments of the *a* values to the various hydrogen atoms. As indicated earlier, the data reported in *Landolt–Bornstein* have been critically evaluated, and in our case there is a footnote to one of the five different methods of generating the anion radical, '...results reported seem to be in error'.

6.5.2. IR spectra

There are many collections and compilations of IR spectra. Some of them are bound volumes, others are loose-leaf sheets to which more sheets are added from time to time. How does one know which compilation contains the desired data? One should use the latest edition of the *IR Spectral Index* and its updated supplements published by ASTM. This is the most comprehensive IR index and contains well over 150,000 published IR spectra. It contains indexes of all the major IR collections and compilations [e.g. *Aldrich Library of IR Spectra, American Petroleum Institute Project 44 (APIP-44), Coblentz Society IR Spectra Collection, Sadtler IR Spectra Collection, TRC Data Project* (formerly Manufacturing Chemists Association (MCA) Research Project)]. Another general index that can be used is the *Sadtler Research Laboratories Total Spectra Index* and its updated supplements. This is a general index to all Sadtler spectra compilations (IR, UV, NMR, etc.) as well as to a few other spectra collections (e.g. *Coblentz Society IR Spectra Collection*).

6.37

Let us assume one is interested in the IR spectra of 2,4,6-trimethylaniline (**6.37**). If one's library holds the *Sadtler Collection* or the *ASTM Index*, of course, one should use these sources. However, many small libraries cannot afford those comprehensive collections, but probably hold the *Sadtler Handbook of Infrared Spectra* or Mecke's *Infrared Spectra of Selected Compounds*. The *Sadtler Handbook* holds spectra of about 3000 compounds whereas Mecke's collection reports the spectra of about 2000 compounds. They contain the IR spectra of common compounds followed by a set of indexes. The compounds in the *Sadtler Handbook* are arranged according to groups—hydrocarbons, halogenated hydrocarbons, nitrogen compounds, compounds containing silicon, phosphorus, or sulphur, oxygen containing compounds except those that have a carbonyl group, and compounds having a carbonyl group. Using the index of the *Sadtler Handbook* one can locate the spectrum of the desired compound (aniline, 2,4,6-trimethyl-) as spectrum 502.

At this stage one should mention the Coblentz Society's *Deskbook of IR Spectra*. This is a selection of high-quality spectra of common classes of compounds which could be used in learning the interpretation of IR spectrum. It includes an excellent bibliography on texts, techniques, IR correlations, Raman spectra, far-IR spectra, and gas-phase spectram, together with a list of spectra compilations and information about IR data retrieval.

In addition to the *Sadtler IR Spectra Collection* Sadtler publishes special interest IR collections and collections of IR spectra of commercial compounds. These collections vary considerably in size, e.g. 150 spectra of high-resolution spectra of gases and vapours, 6600 spectra of surface-active agents. The complete list of the various spectra compilations can be obtained from Sadtler research Laboratories or from Heyden and Son Ltd., Spectrum House, Hillview Gardens, London NW4 2JQ.

The IR spectra of anthranilic acid hydrazide (**6.38**) is not reported in the *Sadtler Handbook* nor in the Mecke's Collection. Here one has to turn to the *Sadtler Total Spectra Index* (or to the *ASTM Index*). The indexes are by name, formula, chemical class, and Sadtler spectrum number. Using either the Name Index (anthranilic acid hydrazide) or the Formula Index ($C_7H_9N_3O$) one finds the following information: prism 33,680; grating 11,671; UV 14,303; NMR 5505. In addition to the references to the prism and grating IR spectra of the compound, there are also references to the UV and proton NMR spectra of the desired compound. In order to obtain the IR spectra of compound **6.38** one has to turn to spectrum 33,680 in the *Sadtler Standard Prism Spectra* and /or to spectrum 11,671 in the *Sadtler Standard Grating Spectra Collection*.

6.38

6.5.3. Raman spectra

The number of compilations of Raman spectra is very small. One can use the *Sadtler Standard Raman Spectra Collection*, which contains over 5000 spectra. Another source is *APIP-44 Selected Raman Spectral Data* which is a collection of Raman spectra of hydrocarbons followed by the full ex-

120

perimental conditions under which the spectra was obtained. It should be noted that the Comprehensive Index of APIP-44 does not cover recently added spectra. However, the Raman spectra compilation has its own index of compounds arranged according to the usual API classification. This index is updated from time to time.

The *DMS Raman IR Atlas*, edited by B. Schrader and W. Meier, contains Raman and IR spectra of about 1000 compounds. Although most, if not all, Raman spectra could be located using other sources, this collection has the advantage of bringing together IR and Raman spectra. Furthermore, among the various indexes in this compilation there is a Substituent Index–a list of substances arranged according to their most important substituents.

6.5.4. Ultraviolet and visible spectra

Similar to the collections of IR spectra there are many collections of UV and visible spectra. Some of these collections are produced by the producers of some of the IR spectra compilations which were discussed earlier— *APIP-44* and *TRC Selected UV Special Data, Sadtler Standard Ultraviolet Spectra, DMS UV Atlas of Organic Compounds*, etc.

A good source which should be held by any library is the *Sadtler Handbook of UV Spectra*. It contains spectra of about 2000 compounds having maxima above 215 nm and a molar absorptivity of more than 100 mol^{-1} cm^{-1}.

Where could one find the UV spectrum of $2H$-1-benzopyran-2-one,8-(2,2-diphenylethyl)-7-methoxy (**6.39**). The spectrum could not be located in any of the sources mentioned earlier (including *Beilstein*). Is there any comprehensive source that could be used, or perhaps one should turn to the *CA*?

6.39

The most comprehensive collection of UV and visible spectra is *Organic Electronic Spectral Data*, published by Wiley–Interscience. This is a collection of all the UV and visible spectra of organic compounds reported in the literature since 1946. After publishing the first two volumes, which cover the literature from 1946 to 1955, a regular publication schedule was worked out. First the average coverage period of each volume was two years. Later, starting in 1966, each volume covers a one year period. However, there is a large publication gap (at present it is 5–6 years) and the volume covering the 1980 literature was published only in 1986. Throughout the series, data are reported from well over 140 journals. Some of the journals used in the early years of the series were omitted later as they no longer contain useful data, and others were added during the years. Altogether it is a compilation of UV spectra of well over 450,000 compounds. The compounds are arranged according to their formulae using the Hill system. Each compound has its name, solvent used (or phase), wavelength, maxima, shoulders and/or inflections, followed by a reference (number of journal, page, and year). The great disadvantage of this series is that there is no cumulative index or list of compounds; thus one has to check the source volume by volume in order to locate the compound one is seeking.

Before starting to look for the spectrum of the compound **6.39** one has to calculate its molecular formula, which is $C_{24}H_{20}O_3$. Starting from the latest available volume (Volume 22—1980) one finds after searching nine volumes the desired information. In Volume 14 (which covers the year 1972) one finds the following information under $C_{24}H_{20}O_3$: 2H-1-benzopyran-2-one, 8-(2,2-diphenylethyl)-7-methoxy-,EtOH, 250 (3,6), 259 (3,7), 323 (4,1), 2-0567-72, and EtOH, 250 (3,3), 325 (3,7), 2-0567-72. The meaning of the data was not clear to our chemist and he decided to check the original source [*Ind. J. Chem.*, 567, (1972)] (whenever possible one should examine the original source). Here log epsilon of compound **6.39** in ethanol was reported to be 3.3 and 3.7 at 259 and 325 nm, respectively.

The search was long, and it would take about the same time (or maybe somewhat shorter) to search manually in the *CA* for compound **6.39** (an online search would take a minute or two) and to find out that it was reported only once in the *CA*—78:29547e [*Indian J. Chem.*, **10**, 567, (1972)]; examination of the original work would give the desired spectral data.

Another useful source (which gives complete drawn spectra) is the series *Absorption Spectra in the Ultraviolet and Visible Region*, edited by L. Lang and published by the Akademia Kiado, Budapest. The first 20 volumes, which were published in cooperation with Academic Press, do not contain volume indexes; however, there are cumulative indexes for Volumes 1–5, 6–10, 11–15, and 16–20. Starting from Volume 21 publication is in cooperation with Krieger, Basel, Switzerland, and each volume has its own index.

6.5.5. Nuclear Magnetic Resonance (NMR) spectra

6.5.5.1. Proton Nuclear Magnetic Resonance (^1H NMR) spectra

Only a few ^1H NMR spectra compilations exist, including the *API Collection*, the *TRC Collection*, and the *Sadtler Standard Nuclear Magnetic Resonance*

Spectra. The last is the most comprehensive collection of ^{1}H NMR spectra, totalling over 35,000 spectra. One could locate a desired spectrum in this collection either by using its latest cumulative index or using the *Sadtler Total Spectra Index*.

About 3000 of the spectra are reproduced in the *Sadtler Handbook of NMR Spectra*, which is a very useful handbook for everyone's usage. Some 480 selected spectra are given in W.W Simon's book, *The Sadtler Guide to NMR Spectra* to illustrate the interpretation of the spectra of the various functional groups. A similar book is N.F. Chamberlain's *The Practice of NMR Spectroscopy with Spectra-Structure Correlation for Hydrogen*.

Unfortunately, there is no general index to ^{1}H NMR; the only available index is *NMR Data Tables for Organic Compounds*, edited by F.A. Bovey, covering the period up to the end of 1962.

In most cases one would end up searching for ^{1}H NMR in the *CA*; the quality of the searching results depend entirely on the quality of the data reported in the literature. Searching for the ^{1}H NMR of ethanone, 1,2,2-triphenyl- (**6.40**) results in the retrieval of a very good spectral data which was reported by G. Kobrich and J. Grosser in *Ber.*, **106**, 2626 (1973).

6.40

6.5.5.2. ^{13}C *Nuclear Magnetic Resonance (*^{13}C *NMR) Spectra*

There are three compilations of 13C NMR spectra. The *Sadtler Standard* ^{13}C *NMR Spectra* which contains about 15,000 spectra, ^{13}C *NMR Spectral Data*, edited by W. Bremser, L. Ernst, B. Franke, R. Gerhards, and A. Hardt, and *Atlas of* ^{13}C *NMR Data*, edited by E. Breitmaier, G. Hass, and W. Voelter. The ^{13}C *NMR Spectral Data* contains over 45,000 spectra of about 40,000 compounds. It contains a small introductory and instructional booklet and a collection of about 260 COM frames (the equivalent of over 54,000 pages of computer output). There are ten indexes in order to enable the user to achieve rapid access to the spectral information: Chemical Name, Formula, Molecular Weight, CAS Registry Number, Author (literature references, only first author), Chemical Shifts, Substance Codes, Structure Coupling Constants, Coupling Constants vs. Structure, and Relaxation Time Indexes.

Similarly to *Sadtler Handbook of NMR Spectra* which deals with ¹H NMR Sadtler published the *Sadtler Guide to* ¹³*C NMR Spectra* which is a useful guide to new as well as old timers. Those interested in the ¹³C NMR (and/or ¹H NMR) spectra of polymers should use *Proton and Carbon NMR Spectra of Polymers* by Q.T. Pham, R. Petiaud, and M. Waten; it is a three volume book published by Wiley.

There are two online data bases devoted to ¹³C NMR, both of which do not have a printed product. One of the data bases is the ¹³C NMR Data Base (C13 NMR) which is produced by BASF AG and at present contains data on about 55,000 ¹³C NMR spectra including relaxation times, coupling constants, and the molecular and structural formulae, of the various compounds. Unfortunately, this data base does not contain at present CAS RNs and retrieval of data cannot be carried out by using the CAS RN. The other one is the CNMR (¹³C NMR Search System) which is produced by the Netherland Information Combine; it contains about 10,000 spectra including relaxation times, coupling constants, molecular and structural formulae, and CAS RNs. It is an integral part of Chemical Information System (CIS) and of Information Consultants Inc. (ICI)

Retrieval of ¹³C NMR spectra of 2-methylpyridine (**6.41**) in the ¹³C NMR Data Base yield nine different spectra, preceded by their bibliographic references. The same search in the CNMR yield two spectra (those two were among the nine retrieved from the ¹³C NMR Data Base).

6.41

However, there are many cases in which the information needed is only partially reported in the literature. The ¹³C NMR of **6.40** is not reported in any of the compilations and/or online data bases reported earlier. Searching the *CA* one comes out with a bibliographic citation [S. Bradamante and G.A. Pagani, *J. Org. Chem.*, **45** , 105 (1980)]. Indeed the Bradamante paper gives the chemical shifts of the various carbon atoms of the molecule except one carbon atom — the carbonyl carbon. If we would like to know the chemical shift of the carbonyl carbon we would have to obtain or make a small amount of compound **6.40** and run the measurements ourselves.

6.5.6. Mass spectra

There are a few compilations and collections of mass spectra (e.g. *APIP-44 Selected Mass Spectral Data*, the *ASTM Index of Mass Spectral Data*). However, the three most useful compilations are A. Cornu and R. Massot's *Compilation of Mass Spectral Data*, S.R. Heller and G.W.A Milne's *EPA/NIH*

Mass Spectral Data Base, and the third edition of the *Eight Peak Index of Mass Spectra*.

The *Compilation of Mass Spectral Data* was published in two volumes by Heyden in 1975 and contains spectral data of about 10,000 compounds. The data are reported in a table giving the ten strongest peaks of the compound. It is arranged by increasing molecular weight, by formula, and by increasing fragment/ion (*m/e*) values. The arrangement of data by *m/e* values enables the searcher to locate a compound on the basis of its ten strongest peaks. The *EPA/NIH Mass Spectral Data Base* is a collection of over 42,000 spectra arranged in order of increasing molecular weight, followed by a set of indexes (Substance, Formula, Molecular Weight, and CAS RN Indexes). The third edition of the *Eight Peak Index of Mass Spectra* contains the eight most abundant ions in 66,720 mass spectra of 52,350 compounds. It is compiled and published by the Mass Spectrometry Data Centre at the University of Nottingham.

A mass spectral compilation which is important not only to chemists but to the health profesionals is the *Mass Spectral and GC Data of Drugs, Poisons and Their Metabolites*, edited by K. Pfleger, H. Maurer, and E. Weber and published in two volumes by VCH Pubs.

How could one find the mass spectra of phenoxyacetic acid (**6.42**) and of ferrocene pentafluorobenzoyl (**6.43**)?

6.42

6.43

The empirical formula of **6.42** is $C_8H_8O_3$ and its molecular weight is 152; one can locate the desired data in the *Compilation of Mass Spectral Data* under either parameter. In both cases one would obtain the following data:

Ten strongest peaks: 107 77 152 94 79 51 65 39 78 108
Relative abundance: 1000 929 812 242 210 208 149 121 98 78

The empirical formula of **6.43** is $C_{17}H_{10}F_5FeO$ and its molecular weight is 380. The spectra of this compound could not located in the *Compilation of Mass Spectral Data*. It is just not reported there. However, the desired data on compounds **6.42** and **6.43** could be retrieved from the *EPA/NIH Mass Spectral Data Base*, either by using the Substance Index (phenoxyacetic acid; ferrocene pentafluorobenzoyl), the Formula Index ($C_8H_8O_3$; $C_{17}H_{10}F_5FeO$), the Molecular Weight Index (152; 380), or the CAS RN Index (122-59-8; 31903-78-3). Using any of the above indexes enables one to locate easily the desired spectras (pages 478 and 2968, respectively). In fact, one does not have to use any index at all; the desired compound can be located by scanning the pages giving the spectra of the various compounds having a molecular weight of 152 and 380, respectively (the main work first followed by the supplements if needed). The printed version of the *EPA/NIH Mass Spectral Data Base* gives the desired data only in a graphic form.

There are two mass spectra online data bases whose tapes are also available by lease or licence. One is the computerized version of the *EPA/NIH Mass Spectral Data Base* (better known as MSSS). The other one is *Wiley/NBS Mass Spectral Database*; it contains 80,000 different spectra of about 68,000 compounds. Those two data bases are also available online as part of CIS (the *EPA/NIH Mass Spectral Data Base* is also available on ICI).

Recently a PC version of the *NBS/EPA/MSDC Mass Spectral Data Base* (which was originally known as the *NIH/EPA Mass Spectral Data Base*) became available. It contains 44,000 mass spectra, various 'index' files for rapid data retrieval as well as the desired software for searching the data base. One can locate the desired spectrum by searching the data base for a chemical name, CAS RN, empirical formula or molecular weight. Various options and constraints may be used for more sophisticated searches detailing the presence (or the absence) of certain peaks.

The data base requires an IBM-AT, PS/2 or compatible (although an XT-class is acceptable). Graphic display is optional and EGA, CGA or Hercules graphic boards are needed. A storage of 8 to 14 megabytes on a hard disk (depending on the number of search options which are installed) is also required.

The PC version of the data base differs from the tape version in that some spectra have been replaced by better quality ones of the same compounds; 215 spectra which were found not to be identified by a recognized chemical name and valid (or correct) CAS RN have been deleted.

How could one run an online search on MSSS in order to obtain the mass spectra of AMP **(6.44)** (better known as 5-adenylic acid). The easiest way would be to search with the CAS RN (the CAS RN of compound **6.44** is 61-19-8), however, searching in this way gives a negative result—no spectra is recorded for a compound with a CAS RN of 61-19-8. This looks strange to our chemist but he decided instead to use the *CA*. Indeed, an online search of the *CA* gave him within about four minutes a list of 18 papers dealing with the mass spectra of AMP.

6.44

⇐ FILE CA
FILE 'CA' ENTERED AT 02:12:30 ON 18 JUL 86
COPYRIGHT 1986 BY THE AMERICAN CHEMICAL SOCIETY

⇐ S 61-19-8(L)(MS OR MASS(W)SPECTR?) ⟵——————————(1)
    7521 61-19-8
     1900 MS
    86769 MASS
   443260 SPECTR?
L1   18 61-19-8(L)(MS OR MASS(W)SPECTR?) ⟵——————(2)

⇐ D L1 1-18 ⟵————————————————————————————(3)

L1   ANSWER 1 OF 18

AN   CA104(14):119198b
    *
     *
      *
       *

⇐ LOG Y
STN INTERNATIONAL LOGOFF AT 02:16:39 ON 18 JUL 86

(1) The L operator is used as the mass spectra term should occur in the text modification which accompanies the CAS RN of AMP (61-19-8). Two terms are used accompanying the CAS RN using the operator OR: ms which is an abbreviation of mass spectra and mass(w)spectr?. ? is a function symbol standing for any number of characters, including none (e.g. spectra, spectral, spectrum). The w operator is used for search for two adjacent terms in the specified order—mass spectr?.
(2) The system finds 18 relevant answers and places them in answer set L1.
(3) The Display command is used to display all 18 answers (1-18) in the bibliographic format.

ₚ ₚ ₐₑₙₒₛᵢₙₑ [ᵤₑ⁻ᵤₗ⁻]*
Adenosine 3',5'-cyclic monophosphate trime≏
thylsilyl derivatives [32645-64-0]   5210
Adenyldeoxyriboside [958-09-8]   1764
3'-Adenylic acid, N-(trimethylsilyl)-2',5'-bis-≏
O-(trimethylsilyl)-, bis(trimethylsilyl) es≏
ter [32653-17-1]   5247
5'-Adenylic acid, pentakis(trimethylsilyl) deriv.
[56145-14-3]   3870 ⟵
5'-Adenylic acid, pentakis(trimethylsilyl) deriv.
[56247-35-9]   3870 ⟵
5'-Adenylic acid, N-(trimethylsilyl)-2',3'-bis-≏
O-(trimethylsilyl)-, bis(trimethylsilyl) es≏
ter [32653-14-8]   5247 ⟵
Adepress [50-48-6]   2024
ₐ ₔₙᵣᵢₗ ᵢ ₅ₙ ₔ ₅ᵢ   ₙₙₙ ₄

Figure 6.2   Substance Name Index of the *EPA/NIH Mass Spectral Data Base*

Me_3SiNH

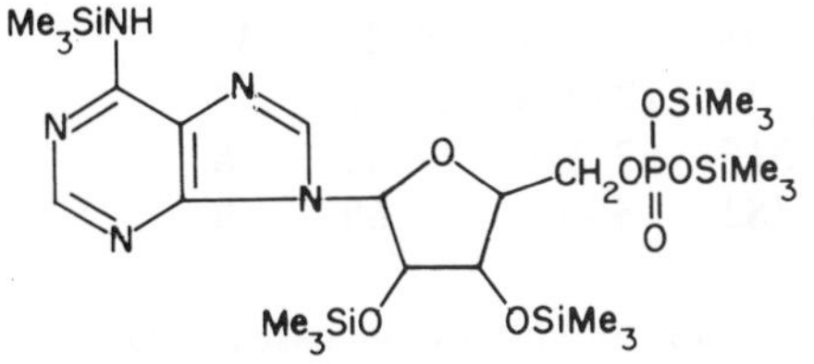

6.45

Our chemist decided to check the Cumulative Substance Index of the *EPA/NIH Mass Spectral Data Base*; he did not find any entry under 5-adenylic acid but there were a few entries to 5-adenylic acid derivatives (Fig. 6.2). Indeed, these were the mass spectra of AMP and were among the ones retrieved from the *CA*. However, the data base structure does not allow the retrieval of the parent compound; only the exact derivative could be retrieved. The data could be obtained from the *MSSS* online by using any of the RNs of the three AMP derivatives, contrary to the printed version where data are available only in graphic form the online version displays the data on both graphic and table (numeric) format. The retrieval of the MS of compound **6.45** is given below:

Options? MSHOW
Enter display mode (TYPE, PLOT, or DPLOT) : TYPE

Enter File number or CAS RN for display (CR to exit): 32653-14-8

CAS RN	QI	NW Formula, Names
32653-14-8	384	707 C25H54N5O7PSi5
		5'-Adenylic acid, N-(trimethylsilyl)-2',3'-bi
		s-O-(trimethylsilyl)-, bis(trimethylsily

128

l) ester (9CI)
Adenosine, N-(trimethylsilyl)-2',3'
-bis-O-(trimethylsilyl)-, 5'-[bis
(trimethylsilyl) phosphate] (8CI)

Instrument: ; Inlet: ; Source temp.: O C; ev: O
Contributor : : JACSAT,90-4183-1B
There are 117 peaks in this spectrum. List (Y/N) ? Y

User: the min & max m/z are: 50 750

User: type all m/z's with intensity > 5

m/z	INT	m/z	INT	m/z	INT	m/z	INT	m/z	INT
73	100	74	9	75	11	129	5	133	5
147	15	169	68	170	10	171	5	192	12
206	5	207	5	208	7	211	5	230	36
231	8	236	10	243	12	258	22	259	7
280	5	299	16	315	67	316	17	317	8
337	5	382	14	383	5	466	5	692	8

One should remember that in many cases data are not reported in the original article although it has been measured. Such a case could be seen searching for the mass spectra of compound 6.40. Search results were negative and no data could be obtained using the various numerical and/or bibliographic data bases. A last attempt was made to examine the various papers which report the synthesis of 6.40. Searching for some spectral data one can assume that the data will be reported in a work describing the preparation of the compound. The synthesis of 6.40 has been reported in the literature 30 times during the years 1967–85. Examination of the above 30 papers results in locating a footnote in one of them [J. Amer. Chem. Soc., 95, 3429 (1973)]: '...benzohydrylphenylketone (8) fragments in the mass spectrometer (70 eV) to give major peaks at m/e 167 ($C_6H_5CHC_6H_5$) and 105 ($C_6H_5CO^+$). In the mass spectrum of 8 obtained from benzene d_6 trapping only the former peak appears at higher m/e'. It is clear that mass spectra measurements of compound 6.40 were made but they were never reported in the literature. Again the solution is to get a sample of the compound and run the spectra. Sometimes one wonders if one should try to retrieve the numeric data that is needed or whether it would be better to carry out the measurements oneself to begin with.

There is an abstract service devoted entirely to mass spectrometry which could also be used for locating numerical data. It is available as paper product and in computerized form: the Mass Spectrometry Bulletin published monthly by the Royal Society of Chemistry (available online via ESA-IRS).

6.5.7. Microwave spectra

The only available compilation of microwave spectra is the *Microwave Spectral Tables* published during the 1960s by the NBS (NBS Monograph No. 70). The compilation is divided into five volumes: Diatomic Molecules, Line Strengths of Asymmetric Rotors, Polyatomic Molecules with Internal Rotation, Polyatomic Molecules without Internal Rotation, and Spectral Line Listing. The last volume lists all of the lines which have been reported in the previous volumes in ascending frequency. Thus it could be used as a ready reference for unknown spectral lines.

Critical reviews on molecular-rotational spectra in the microwave frequency range which give corrected and up-to-date data are the NBS publications which were published in the *Journal of Physical and Chemical Reference Data*: F.J. Lovas and E. Tiemann, 'Microwave Spectral Tables, I. Diatomic Molecules', *J. Phys. Chem. Ref. Data*, **3**, 609 (1974), which corresponds to Volume I of the NBS Monograph No. 70 and covers the published literature up to June 1973, and F.J. Lovas, 'Microwave Spectral Tables, II. Triatomic Molecules', *J. Phys. Chem. Ref. Data*, **7**, 1445 (1978), which corresponds to Volume IV of the NBS Monograph No. 70 and covers the published literature up to 1976.

A collection of papers dealing with the details of microwave spectra of over 25 compounds of astrophysical interest (e.g. formaldehyde, methylamine, methanol, acetaldehyde, acrylnitrile, methyl formate) have been published by scientists from the Molecular Spectra Data Center of the NBS in the *Journal of Physical and Chemical Reference Data*. The full bibliographic information can be found at the end of the latest issue of this journal.

6.5.8. Mossbauer Spectroscopy

Numerical and bibliographical data as well as abstracts of the literature on the Mossbauer effects are given in the *Mossbauer Effect Data Index*, in the *Mossbauer Effect Reference and Data Journal*, and in the *Mossbauer Spectroscopy Abstracts*. The Index covers the period from 1958 to 1976. The first volume covers the literature for the years 1958–65, the second volume covers the years 1966–68, and the other volumes cover one year each. There is a Cumulative Index to the Mossbauer Effect Data Index, covering the indexes of the ten volumes of the publication. While the first volume was published by Interscience all subsequent volumes as well as the Cumulative Index were published by IFI/Plenum.

Publication of the *Mossbauer Effect Reference and Data Journal* started in 1978. Ten issues are published annually by the Mossbauer Effect Data Centre at the University of North Carolina—Asheville, Asheville, NC 28804, USA. There are sections for each isotope for which relevant data appeared in the coverage period. Abstracts that cannot be classified under specific isotopes are grouped together into sections (e.g. analysis, reviews).

The *Mossbauer Spectroscopy Abstracts* is a quarterly publication covering the various aspects of the Mossbauer spectroscopy. It is a bibliographic source published since 1978 by the PRM Science and Technology Agency Ltd. of London.

6.6. Thermophysical and thermochemical data

6.6.1. Thermophysical data

The most comprehensive collection of evaluated thermophysical data is the *Thermophysical Properties of Matter*, which is produced by the Thermophysical Research Data Centre (TPRC) and published by IFI/Plenum. The TPRC is one of the centres located within CINDAS at Purdue University. The compilation is composed of thirteen volumes and a supplement to Volume 6 (altogether fourteen volumes) which were published during the years 1970–77. There is a Master Index to the series which combines all of the indexes of the individual volumes. This Master Index should be used when one is looking for thermophysical data of an inorganic compound. When using this source one has to keep in mind that some of the volumes were published in the early 1970s and their coverage period does not extend beyond 1970 or 1971.

Bibliographic citations to thermophysical properties could be located using the *Thermophysical Properties Research Literature Retrieval Guide*. This compilation is composed of the Basic Edition, covering the literature from 1920 to 1963, Supplement I, covering the literature from 1964 to 1970, and Supplement II, covering the period 1971–77. Each Supplement is divided into six retrieval guides according to material class. A Cumulative covering the period 1900–80 has been published as well.

Let us try to find the specific heat of calcium fluoride at low temperatures using the *Retrieval Guide*. Calcium fluoride is an inorganic compound and should be reported in Volume I—Elements and Inorganic Compounds. The first step is to locate the material name in Part A of Volume 1. On p. A9 we locate the following entry—Calcium Fluoride CaF_2—a e g h i j n o 106-0026. There is also a cross-reference to Fluorite 521-0096, which is a mineral form of calcium fluoride. The list of letters which followed the compound name is a list of properties reported for the compound. The code for specific heat (the property we are interested in) is e, thus it is reported in this compilation. The second step is to record the seven digit substance number, 106-0026, and turn to the property chapter. The chapter one should turn to is Specific Heat (chapter 3, p. B25). The seven-digit substance number can be found in p. B36 columns 1 and 2. There are six entries, of which only the first two have the parameter of interest (low temperature). These two entries report TPRC numbers—56771 and 75410. In order to locate the desired reference one has to turn to the bibliography part. There one can locate on p. C17, colunm 1 and on p. C207, column

1, the two references. The desired numerical data can be obtained from the original literature.

In 1981 the CINDAS made available for lease or licence the Thermophysical Properties of Materials: Computer Readable Bibliographic Files. It is a magnetic tape containing the CINDAS files of thermophysical properties (having 62,000 references to 44,400 compounds) as well as electronic, magnetic, and optical properties (having 22,000 references to 13,100 compounds). The tape in updated quarterly.

6.6.2. Thermochemical and thermodynamic data

The compilation and evaluation of numerical data is a slow and painstaking job. Some of the data compiled might be old by publication time, and duplication in compilation might also occur. How can one find out about data compilation projects in progress, unpublished complete reports and numerical data reported in the current literature? This is a serious problem with any kind of numerical data, but in the case of thermochemical and thermodynamic data there is a source which gives information about ongoing projects and research, results of projects finished but not yet published, and recent data reported in the literature - it is an annual publication entitled *Bulletin of Chemical Thermodynamics* (formerly the *Bulletin of Thermodynamics and Thermochemistry*).

Where could a chemist find the standard enthalpy of formation ($\Delta H_f°$) and the entropy ($S°$) at 298.15 K of UCl_3? The first place to look in is the third edition of the *JANAF Thermochemical Tables*. The *Tables*, which were compiled by the Dow Thermal Research Laboratory staff and published by the NBS gives the thermodynamic properties (heat capacity, entropy, Gibbs energy function, enthalpy, heat of formation, and Gibbs energy of formation) in the solid, liquid, and/or ideal gaseous state over the temperature range of 0–6000 K. The following elements and some of their simple compounds (those with halogens, oxygen, and hydrogen) are covered Al, Ar, B, Ba, Be, Br, C, Ca, Cl, Co, Cr, Cs, Cu, D, F, Fe, H, He, Hg, I, K, Kr, Li, Mg, Mo, N, Na, Nb, Ne, Ni, O, P, Pb, S, Si, Sr, Ta, Ti, V, W, Xe, and Zr. The third edition of the *Tables* have been published in two volumes as supplement 1 to the *Journal of Physical and Chemical Reference Data* Volume 14 (1985).

Other series of thermodynamic tables are the *International Thermodynamic Tables of the Fluid State*, the *Key Chemical Data Books*, and the *API Monograph Series*. The *International Thermodynamic Tables of the Fluid State* is a series of books published by Pergamon Press for IUPAC as part of their Chemical Data Series. Each book gives various evaluated thermodynamic data of an element or a chemical compound. The following elements and compounds are covered: argon, ethylene, carbon dioxide, helium, methane, nitrogen, propylene (propene), fluorine, chlorine. *Key Chemical Data Books* are published by the TRC and give evaluated thermochemical and thermodynamic data on the following organic compounds: phenol, benzene, cresols, xylols, furan, dihydrofuran, and tetrahydrofuran. The *API Monograph Series* which is produced by the API is very similar to the *Key Chemical Books* and covers the following organic compounds: tetralin, *cis-* and *trans-* decatin, naphthalene, anthracene and phenanthracene, four-ring condensed aromatic compounds.

On examining the complete list of the *JANAF Tables* one finds that desired data on UCl_3 are not recorded in the *Tables*.

Another source is I. Barin and O. Knacke's *Thermochemical Properties of Inorganic Substances* which was published in 1973 by Springer-Verlag. A supplement which included data on new compounds and new data on compounds reported in the main volume was published in 1977. The supplement also contains a detailed description of how to use the tables (a similar book of tables for organic compounds is J.D. Cox and G. Pilcher's *Thermochemistry of Organic and Organometallic Compounds*, published in 1970 by Academic Press). Our chemist was not able to get Barin and Knacke's book in his library, and such a reference book usually would not be available for an interlibrary loan. Furthermore, no library in the immediate vicinity was in possession of the book. Under such circumstances it was logical to try to locate another source.

At that stage our chemist decided to check the *Gmelin* (in fact, he should have begun his search using this source). The Uranium Supplement Volume C9 is devoted to halogen compounds. The various physical properties of UCl_3 are reported on pages 4–7, on page 5 one can locate the following information: $\Delta H^\circ_{f298} = -213.5 + 2.0$ kcal mol^{-1} and $S^\circ_{298} = 38 + 0.2$ cal mol^{-1} K^{-1}, followed by a few references. The most important reference was M.H. Rand and O. Kubaschewski— *The Thermochemical Properties of Uranium Compounds*, published in 1963 by Oliver and Boyd (Edinburgh). Another source that could have been used is the *Selected Values of Chemical Thermodynamic Properties*, edited and compiled by D.D. Wagman *et al.* This is a series of eight NBS Technical Notes (270, 1–8) giving enthalpies, free energies of formation, entropies, and heat capacities at 25'C of various compounds of the elements. A single volume version of those Technical Notes entitled *NBS Tables of Chemical Thermodynamic Properties* has been published in 1982 as Supplement 2 to *J. Phys. Chem. Ref. Data*, **11** (1982). The Technical Notes or the Tables are a revision of the first part of F.D. Rossini *et al.*, *Selected Values of Chemical Thermodynamic Properties* (NBS Circular 500), which was published in 1952. The second part of Circular 500 reports enthalpies, entropies, and heat capacities related to phase transitions of the elements, inorganic compounds, and organic compounds having one or two carbon atoms.

VINITI produced its series *Thermal Constants of Substances*; it is very similar in its coverage to the NBS Technical Notes except that it also gives data for properties that characterize phase transition (T_{trs}, $\Delta_{trs}H$) and it also contains organic compounds with one or two carbon atoms.

A source for evaluated chemical thermodynamic compilation of biological and/or aqueous systems is G.T. Armstrong and R.N. Goldberg's *Annotated Bibliography of Compiled Thermodynamic Data Sources for Biochemical and Aqueous Systems (1930–1975)* (NBS Special Publication No. 454).

6.7. Kinetic data

Let us consider a physical chemist who is interested in obtaining the stability constant of the complex of Ni^{2+} with the dipeptide D-leucyl-L-tyrosine, the rate constant of the base-catalysed hydrolysis of isopropyl acetate (**6.46**) and the rate constant of the gas-phase decomposition of the ester **6.46** to acetic acid and propylene. Where could he locate these constants?

$$CH_3COOCH\overset{\displaystyle CH_3}{\underset{\displaystyle CH_3}{\Big\langle}}$$

6.46

One source that can (and should) be used to look for bibliographic and tabular sources for kinetic data is Chapter 6 of the *CODATA Directory*. This chapter which was edited by L.H. Gevantman and published as *Bulletin* No. 46 in July 1981 should list, among other subjects, sources for stability constants of metal ion complexes, reaction rate constants, etc.

6.7.1. Stability constants of metal ion complexes

Stability constants of metal ion complexes can be located using *Stability Constants of Metal Ion Complexes*, compiled by L.E. Sillen (inorganic ligands) and A.E. Martell (Organic ligands) and published by the Chemical Society (Special Publication No. 17), which is a collection of metal complexes equilibrium constants. Two supplements to this compilation have been published. The first covers the period 1963–69 and was published by the Chemical Society (Special Publication No. 25); the second covers the period 1970–73 and was published by Pergamon Press in two volumes: *Stability Constants of Metal Ion Complexes*, Part A, *Inorganic Ligands*, compiled by E. Hogfeldt; *Stability Constants of Metal Ion Complexes*, Part B, *Organic Ligands*, compiled by D.D. Perrin. The equilibrium constants given in the above compilation are unevaluated, so there are cases in which there are differences in a particular equilibrium constant of an order of magnitude based on two different sources. Evaluated data are given for some of the equilibrium constants in *Critical Stability Constants*, compiled by A.E. Martell and R.M. Smith and published by IFI/Plenum as a four-volume compilation (Amino Acids, Amines, Other Organic Ligands, Inorganic Ligands). A First Supplement to this compendium covering the literature through to 1979 has been published in 1982.

134

Stability constants of log $K = 3.73$ and 6.66 were found for Reactions 6.4 and 6.5, respectively, using the original compilation of *Stability Constants of Metal Ion Complexes* (compound 1025 on page 763). The same data was found in Volume 1 of *Critical Stability Constants* (log $K = 3.59 \pm 0.1$ and 6.53 ± 0.1, respectively).

$$Ni^{2+} + HL \longrightarrow NiHL^{2+} \qquad (6.4)$$

$$Ni^{2+} + 2HL \longrightarrow NiHL^{2+} \qquad (6.5)$$

$$L = D-leucyl-L-tyrosine$$

6.7.2. Reaction rate constants

Our physical chemist also needs the rate constant of the base-catalysed hydrolysis of isopropyl acetate (**6.46**) to acetic acid and 2-propanol (Reaction 6.6) and the rate constant of the gas-phase decomposition of the ester **6.46** to acetic acid and propene (Reaction 6.7).

$$CH_3COOCH(CH_3)_2 \longrightarrow CH_3COOH + CH_3CHOHCH_3 \quad (6.6)$$
$$\mathbf{6.46}$$

$$CH_3COOCH(CH_3)_2 \longrightarrow CH_3COOH + CH_2=CHCH_3 \quad (6.7)$$

Our chemist knew about the NBS *Tables of Chemical Kinetics, Homogeneous Reactions* and its various supplements (the original *Tables* were published in 1951 as Circular 510 followed by two Supplements published in 1961 and 1964 as Monographs 34 and 34-2, respectively). This set of tables suffers from not having been updated for well over 20 years. However, they contain valuable data which could not easily be obtained otherwise. The tables are arranged in numerical order, each table has a six-digit number, the first two digits describing the reaction type, the third digit the reaction phase (gases—1; liquids—2; solids—3), and the last three digits the types of substances involved in the reaction.

Using the Subject Index our chemist found that the original *Tables* (Circular 510) and the First Supplement contain Table 212.441—'Ester Solvolysis of Aliphatic Carboxylic Acids and Alcohols'. The hydrolysis reaction our chemist is interested in—Reaction 6.6, is not given in the table in the original compilation but in the First Supplement it is reported on page 101. The rate constant (L mol^{-1} sec^{-1}) is given for three different temperatures:

1.05 × 10^{-3} at 0 °C, 2.33 × 10^{-3} at 10 °C, and 4.82 × 10^{-3} at 20 °C. The measurements were carried out in 62 wt.-% acetone − 38 wt.-% water (*ca.* 67 vol.-% acetone − 33% vol.-% water). The data were followed by a reference to the original literature [*J. Am. Chem. Soc.*, **72**, 3021 (1950)]. No data could be obtained using these *Tables* concerning the gas-phase decomposition of the ester.

In order to obtain more recent data about the base-catalysed hydrolysis of the ester our chemist turned to E.T. Denisov's compilation *Liquid Phase Reaction Rate Constants*. He found that this compilation is not the one he thought it was; it deals only with radical reactions: monomolecular reactions in which radicals are formed due to decomposition of the molecule, formation of free radicals in bi- and trimolecular reactions, radical reactions, and ion-molecule reactions that are accompanied by the formation of free radicals. No relevant data could be found in this source.

6.7.2.1. Comprehensive Chemical Kinetics

Comprehensive Chemical Kinetics edited by C.H. Bamford, C.H.F. Tippet, and R.G. Compton is a multi-volume publication (of over 25 volumes) in ten sections: Practice and Theory of Kinetics, Homogeneous Decomposition and Isomerization Reactions, Inorganic Reactions, Organic Reactions, Polymerization Reactions, Oxidation and Combustion Reactions, Selected Elementary Reactions, Heterogeneous Reactions, Kinetics and Technological Processes, and Modern Methods − Theory and Methods.

Information about the base-catalysed hydrolysis of isopropyl acetate should be found in Volume 10, *Ester Formation and Hydrolysis and Related Reactions*. Indeed, in this Volume (in Chapter 2) there is a table (Table 31) entitled 'Hydroxide Ion Catalysis'. Our chemist was able to find there a rate constant for the reaction which was measured in 70% acetone at 24.4 °C. The value given was 7.06 × 10^{-3} (L mol^{-1} sec^{-1}), which did not seem to agree with the values he had already found (1.05 × 10^{-3} at 0 °C, 2.33 × 10^{-3} at 10 °C and 4.82 × 10^{-3} at 20 °C). As the data were accompanied by the literature reference [*J. Chem. Soc. B*, 661 (1966)] he realized that he should turn to the original literature (and may be also to look for more recent data) in order to evaluate the quality of the data he had found.

In order to obtain the rate constant of the ester decomposition (Reaction 6.7) one has to use Volume 5 of the treatise *Decomposition and Isomerization of Organic Compounds*. In it is a chapter entitled 'The Unimolecular Decomposition and Isomerization of Oxygenated Organic Compounds'. Two subsections of the chapter seem to be relevant: '2.1.2 Acetate Esters' and '2.1.3 Substituent Effects'. Indeed in subsection 2.1.3 there is a long table in which one could locate the relevant data followed by the original references (Table 6.7).

Table 6.7   Data on the decomposition of isopropyl acetate

Log A	F_a	ΔT (K)	Reference
13.0	45.0	715–801	*Can.J.Chem.*, **32**, 366 (1954)
13.38	46.6	650–710	*Recl.Trav.Chim.Pays-Bas*, **82**, 1123 (1963)
13.4	46.34	586–635	*J.Chem.Soc.*, 335 (1962).

Using these data our chemist could easily calculate the rate constant of the reaction.

6.7.3. Data evaluation

As our chemist was not happy with the hydrolysis rate constants he found earlier he decided to try to evaluate the data. To begin with he decided to retrieve any relevant information on the rate of basic hydrolysis of isopropyl acetate from 1967 onwards (the last reference he had dated back to 1966). Indeed a search in *CA* results in 15 relevant references. However, only 5 of those 15 he could examine in his university library; the others were not available in his university nor in nearby libraries. Even the five articles that he was able to read made his attempts more complicated. Some of the rate constants are given below: in 90% aqueous p-dioxane log K was 0.1267 L mol^{-1} min^{-1} at 20 °C, and 0.6175 L mol^{-1} min^{-1} at 50 °C, in water K was found to be 1.28–1.29 L mol^{-1} min^{-1} at 25 °C, 1.58–1.72 L mol^{-1} min^{-1} at 25 °C and 3.37–3.42 L mol^{-1} min^{-1} at 30 °C; in 50% aqueous p-dioxane K was found to be 1.049 L mol^{-1} min^{-1} at 30 °C and 1.855 L mol^{-1} min^{-1} at 40 °C; in 70% aqueous acetone K was found to be 0.5649 L mol^{-1} min^{-1} at 30 °C, 0.709 L mol^{-1} min^{-1} at 35 °C and 0.901 L mol^{-1} min^{-1} at 40 °C. One can see that some work should be done in order to evaluate the available numeric data.

Critical evaluation of these data would have to be carried out afresh every time anyone is seeking this information. One may look at it as making nonsense of the concept of economy of effort. Thus it would be only reasonable, meaningful, and rational not to leave it to each individual end user to evaluate the data, but instead to make the information available after critical evaluation by a 'data analysis centre'. This demands the creation of many more data analysis centres. It also means that a user looking for any kind of numerical data should examine first whether the data he is after had been screened and/or evaluated by a data analysis centre.

6.8.   Solubility data

Although qualitative or semiquantitative solubility data can be located easily (in any chemical handbook or dictionary), quantitative data are harder to obtain. Quantitative data for the solubility in water of a limited number of organic and inorganic compounds can be found in R.K. Freier's book *Aqueous Solutions*, a two-volume compilation published by Walter de Gruyter. The first volume contains solubility data in water (as well as dissociation constants and redox potentials) of about 4000 inorganic and organic compounds. The second volume contains additional information about selected compounds, over a third of it being devoted to water and heavy water (H_2O and D_2O).

Some solubility data could be found in the following volumes of *Landolt–Bornstein*: II/2b (of the sixth edition), *Solution Equilibria*; IV/4c (of the sixth edition), *Absorption Equilibrium of Gases in Liquids*; III/3 and III/16 (of the New Series), *Liquid Systems and Their Heat Capacities*.

A much more general compilation is *Solubility of Inorganic and Organic Compounds*, which is an English translation of a Russian compilation, edited by H. Stephen and T. Stephen and published by Pergamon Press. The first two volumes (in four parts) were originally published in Russia in the 1960s and the third volume (in three parts) was published 15 years later. The first volume deals with binary systems and the other volumes are devoted to ternary and multi-component systems.

Bibliographic information about the solubility of gases and solids could be found in a new reference work entitled *Solubility of Gases and Solids: A Literature Source Book*. It was edited by J. Wisniak and M. Herskowitz and published by Elsevier.

Alloys and their phase diagrams are reported and evaluated in many sources. The classical source is M. Hansen and K. Anderko, *Constitution of Binary Alloys*, published by McGraw-Hill, New York, in 1958. Two supplements to this compilation were published in 1965 and 1969, respectively: R.P. Elliot, *Constitution of Binary Alloys* and F.A. Shunk, *Constitution of Binary Alloys*. Other sources one can use are the following:

(1) *Handbook of Binary Metallic Systems—Structure and Properties*, edited by A.E. Vol. Five Volumes of this handbook have been published covering the alloys of Ac, Al, Am, B, Ba, Be, and N (Volume 1); Bi, Dy, Fe, Ga, Gd, Ge, H, Hf. Ho, V, and W (Volume 2); Av, In, Ir, Y, and Yb (Volume 3); Ca, Cd, and K (Volume 4); and Co, O, and Si (Volume 5). The handbook is published in Russian by Nauka, Moscow, but there is an English translation of the first two volumes (these can be obtained from NTIS as documents TT66-51149 and TT66-51150, respectively).

(2) *Selected Values of the Thermodynamic Properties of Binary Alloys* by R. Hultgren, P.D. Desai, D.T. Hawkins, M. Gleiser and K.K. Kelley published by the American Society for Metals (1973). This compendium in-

cludes phase diagrams of 340 binary alloy systems evaluated and updated up to 1973 as well as evaluated thermodynamic data for those systems. This book, together with its companion, *Selected Values of Thermodynamic Properties of the Elements*, is a revised and updated version of *Selected Values of Thermodynamic Properties of Metals and Alloys* by R. Hultgren *et al.*, which was published in 1963.

(3) *Handbook of Binary Phase Diagrams* by W.G. Moffatt published by General Electric, Schenectady, New York (1981). This is a four-volume compilation with semi-annual updates.

Crystallographic information about binary and multi-phase intermetallic phases could be found in Pearson's *Handbook of Crystallographic Data for Intermetallic Phases*. The *Handbook* was edited by P. Villars and L.D. Calvert and published by the American Society for Metals (1985).

When one is looking for a phase diagram of a binary alloy system, which of the above sources should one use? Perhaps one should use some other source. There are over 25 different compilations of binary alloy phase diagrams, so how would one find out where the required data are reported?

Before using any of these compilations one should turn to W.G. Moffatt's *Index to Binary Phase Collections*. This is a bibliography published by General Electric in 1980 which is a merged index to 23 different compilations of binary phase diagrams. Additional compilations will be included in the next edition. Another index which might be of use is the *NBS Alloy Data Centre Permuted Material Index*, which gives literature references to the original work on the physical properties of metals and alloys. The properties given are: electron transport properties, magnetic properties, mechanical properties, nuclear and other resonance properties, quantum descriptions of solids, electromagnetic radiation, superconductivity, thermodynamics, and soft X-ray spectroscopy.

Among the various solubility data projects which are in progress one should be aware of one in particular—the IUPAC Solubility Data Project (about 25 of the planned over 100 volumes of the output of the project have already been published).

The Solubility Data Project is divided into solubility of gases in liquids (data on the solubility of inorganic gases, organic gases, and their mixtures as a function of pressure and temperature in water, aqueous solutions, organic solvents and their mixtures, molten metals, fused salts, ceramics, and glasses; e.g. Volume 1, *Helium and Neon*; Volume 10, *Nitrogen and Air*), solubility of liquids in liquids (data on mutual solubilities and phase formation of binary and multi-component liquid systems at various temperatures and pressures; e.g. Volume 15, *Alcohols with Water*), and solubility of solids in liquids (data on the solubility of various organic and inorganic compounds, metals, alloys, minerals, and rocks in various liquids; e.g. Volume 7, *Silver Chloride in Water*; Volume 16–17, *Antibiotics I, Beta Lactams*; Volume 27, *Metals and Intermetallic Compounds in Mercury*). The data reported are critically evaluated; whenever different data are reported by different groups the evaluator has added a tentative or a recommended solubility table. The data are given in the units used in the original work, but when the data are given in unusual units, they may be recalculated into conventional units. The series will be accompanied by a series of detailed cumulative indexes of which the first one covering Volumes 1-18 has been published in 1985 as Volume 19 of the series.

6.9. Other numerical data

Many other topics of numerical data are of interest to chemists. The first stage in searching numerical data of one type or another is to identify and locate the appropriate source(s) for searching (including the *CA*). Anyone interested in electrochemical data would have to use the *CRC Handbook Series in Inorganic Electrochemistry* (which was published in four volumes by the CRC Press) or the *CRC Handbook Series in Organic Electrochemistry* (which was published in five volumes by the CRC Press). Anyone interested in estimation (or prediction of data) should use the R.C. Ried *et al.* source, *The Properties of Gases and Liquids* (McGraw-Hill, New York, 1977), S.W. Benson's *Thermochemical Kinetics* (Wiley, New York, 1968), or L.E. Nielson's *Predicting the Properties of Mixtures* (M. Dekker, New York, 1978). He may also use one of the predictive modelling and/or simulating online data bases such as CHEMTRAN, DETHERM-SDC, EROICA, F*A*C*T, PPDS, TISDATA, etc.

CHAPTER 7

Chemical General Subjects, Reactions, and Concepts

Information on specific and/or general chemical subjects is required by the chemist in the various stages of his scientific work. Information may be needed as part of a general survey of a new area of research one may be planning to enter or possibly as a background material for a grant application one is writing. It may be needed for decision making in the chemical industry (e.g. patent applications, production of new products). One may need the information in order to interpret some of one's latest experimental results or to clarify and explain problems which arose in the course of scientific work. A chemist may need the information for a lecture he is planning to deliver or for an article he is writing.

Let us examine how one can obtain the desired information that is needed.

7.1. General information

How could a chemistry undergraduate student identify and locate sources for his term paper in nuclear chemistry entitled 'Islands of Stability'?

The first step was to try and locate book(s) where some information is included. Our student found a few sentences on the subject in a nuclear chemistry textbook, but as he needed more information he searched for more specific books. Indeed, in C. Keller's book *The Chemistry of the Transuranium Elements* he found a section dealing with the subject (Section 1.3 — Possibility of Synthesizing New Elements — pp. 27-38). The book was published in 1971 and included references up to 1969. Our student was in need of some recent references on the subject and asked the librarian for help.

The librarian suggested carrying out an online search on the above topic and explained to him that in order to search effectively online he should:

(a) Define his search topic — specifying exactly the purpose of the search as well as the results expected (a comprehensive search, a few good references, etc.).

(b)   Identify the important components of the search topic—pick out the main ideas which must be represented in the search profile.

(c)   Decide on search terms—decide what search terms are needed to represent the ideas of the search topic.

(d)   Construct a search profile—specify the relationships among the various search terms (e.g. Boolean operators, proximity operators, truncations).

(e)   Select the proper file for searching—which data base to search and which system (vendor) to use.

(f)   Conduct the search—instruct the system to do the search.

(g)   Review the results—look at some of the results (usually the titles) in order to determine if he had retrieved the information he was after.

(h)   Modify and/or revise the search profile—according to the review of the results decide if some modification and/or revision in the profile is needed. If so, decide how to modify it in order to obtain the desired results.

(i)   Rerun the search—rerun the modified search. The ability to modify the search profile and then immediately reexecute the search is the highpoint of online searching.

(j)   Print the results of the search—get a hard copy of the final set of the answers, either online or offline.

(k)   Obtain the original documents—determine which of the obtained answers you need to read and get those from your library.

The results of the online search using the *CA* File on STN are shown below:

⇐ FILE CA
FILE 'CA' ENTERED AT 01:17:31 ON 25 JUL 86
COPYRIGHT 1986 BY THE AMERICAN CHEMICAL SOCIETY

⇐ S ISLAND#(L)STABILITY ←——————————————————————————(1)
              5238 ISLAND#
              76210 STABILITY
L1   18 ISLAND#(L)STABILITY

(1) The truncation symbol # is used to indicate zero or one additional character at the end of the word (island, islands). Other truncation symbols used: ? = any number of characters at the end of the word; ! = exactly one character inside or at the end of the word but not at the beginning.

Our student obtained 18 references of which 10 were relevant. The librarian suggested that maybe the same search should be carried out also using a physics data base and explained to the student the difference between *Physics Abstracts* and *Physics Briefs* (*Physikalische Berichte*). The student decided to search both data bases using the same profile search. (One should remember that while the Abstract Field in the *CA* File on STN is not automatically searchable, it is automatically searchable in the various online versions of *Physics Abstracts* (*INSPEC*) and *Physics Briefs*.) Searching Physics Briefs results in obtaining 10 references (8 of them relevant) while the Physics Abstracts search results in 5 references (4 of them relevant).

Altogether our student obtained 18 different relevant references. Only one of the references was retrieved in the three different data bases and two were retrieved in two of the data bases (*CA* and Physics Briefs). One should be aware also that at least one of the references—Aborigines of the Nuclear Desert, A. De Rujula, *Nucl. Phys. A*, **A434**, 605 (1985)—was retrieved only in one data base (*CA*) although it is indexed in all three data bases (*CA* 102:10248a; *Physics Briefs* 85:59917; *INSPEC* A85:52872).

How could an organic chemist locate information about charged (negatively or positively) forms of helicenes and cyclophanes? Our chemist realized that the most important thing he has to do before starting searching is to define and identify the chemical compounds he is interested in. Does he need general information about two classes of positive and/or negative ions (helecenes and cyclophanes) or does he maybe really need detailed information on the various charged species which belong to those classes? What he really wants is detailed information on the various charged species of helicenes and cyclophanes. His next problem was how to identify the various helicenes and cyclophanes. He knew the structures of about 30 different helicenes and cyclophanes; however, it was not clear to him whether this represents 10%, 40%, or maybe 75% of the known defined species of those two classes of compounds. He could identify all chemical compounds having the word helicene or cyclophane in their *CA* Index Name and/or synonym by doing an online dictionary search in the *Registry* (*REG*) File of STN. A crossover of this search results from the *REG* File to the *CA* File would enable him to search for bibliographic citations to his query. The results of the search are given below:

FILE 'REGISTRY' ENTERED AT 04:59:09 ON 17 FEB 86
COPYRIGHT 1986 BY THE AMERICAN CHEMICAL SOCIETY

   ⇐ S CYCLOPHANE OR HELICENE
          254 CYCLOPHANE
          66 HELICENE
L1   320 CYCLOPHANE OR HELICENE

   ⇐ FILE CA
FILE 'CA' ENTERED AT 04:59:46 ON 17 FEB 86
COPYRIGHT 1986 BY THE AMERICAN CHEMICAL SOCIETY

   ⇐ S L1 AND (ION OR CATION OR ANION)
         873 L1
         189712 ION

```
        45404 CATION
        27351 ANION
L2          38 L1 AND (ION OR CATION OR ANION)

⇐ S L1 AND (ION OR CATION OR ANION)/AB
        873 L1
        144559 ION/AB
        41471 CATION/AB
        31914 ANION/AB
L3          38 L1 AND (ION OR CATION OR ANION)/AB

⇐ S L3 NOT L2
L4          18 L3 NOT L2
```

Out of the 38 answers which were obtained searching the Basic Index (L2), 32 were relevant ones. On the other hand, out of the 18 additional answers obtained searching the Abstract Field (L4) only two were relevant. In many cases online searching of the Abstract Field of the *CA* results in getting a small number of new relevant items followed by a large number of irrelevant ones. Indeed, usually an online searching in the *CA* File at STN would not include searching of the Abstract Field. One of the main problems in online searching is the ratio between recall (percentage of relevant items in the File which has been retrieved) and precision (percentage of retrieved items that is relevant to the search topic). Although the ideal situation would be to have high recall and high precision, the real situation is that they usually do not go together—high precision is accompanied by not so high recall and high recall is accompanied by not so high precision.

A pharmaceutical chemist was interested in obtaining bibliographical information about the sustained release of chlorhexidine (**7.1**). When he ran an online search for himself, he was not happy with the results.

$$\text{Cl} \diagdown \bigcirc \text{—NHCNHCNH(CH}_2)_6\text{NHCNHCNH—} \bigcirc \diagup \text{Cl}$$

with $\overset{NH}{\overset{\|}{\text{C}}}\overset{NH}{\overset{\|}{\text{C}}}$ groups

7.1

FILE 'REGISTRY' ENTERED AT 00:20:08 ON 25 JUL 86
COPYRIGHT 1986 BY THE AMERICAN CHEMICAL SOCIETY

```
⇐ S CHLORHEXIDINE/CN
L1          1 CHLORHEXIDINE/CN

⇐ FILE CA
```

144

⇐ S L1(L)(LONG(W)ACTING OR SUSTAINED(W)RELEASE OR CONTROL?(W)RELEASE OR SLOW(W)
RELEASE OR DELIVERY) ⟵————————————————————(1)
   403 L1
  26218 LONG
   2963 ACTING
   2947 SUSTAINED
  55498 RELEASE
 153811 CONTROL?
  55498 RELEASE
   8566 SLOW
  55498 RELEASE
   2806 DELIVERY

L2   1 L1(L)(LONG(W)ACTING OR SUSTAINED(W)RELEASE
OR CONTROL? (W)RE
⇐ D L2
L1 ANSWER 1 OF 1
  *
  *

(1) He included in the search profile various terms which reflect the term sustained release.

  Two references he knew [*J. Periodontal Res.*, **77**, 323 (1982); *J. Peridontal.*, **53**, 693 (1982)] were not retrieved. He prepared his profile based on the knowledge that beginning with the 11th Collective Index period salts of organic acids and bases are indexed as the corresponding free acid or base whenever they were administered to biological systems. Discussing his search and its results with his library information specialist he learned that only inorganic salts of organic acids and bases are indexed, starting at the beginning of the 11th Collective Index period, as the corresponding free parent compounds (the organic acid or base) whenever they were administered to the biological system. However there are exceptions to this—when the compound is described by a trade name, the compound is indexed in the form of the trade name compound; when the compound has been synthesized and characterized as a salt; or when there is a stress on the use of a particular salt form (bioavailability, comparison of activity with other salts of the same acid or base, etc.), in such cases the compound is indexed both as the salt and the parent compound. The information specialist modified somewhat the search profile; an online search of the modified profile gave

145

five answers—the one obtained earlier, the two our pharmaceutical chemist knew, and two new ones.

⇐FILE REG
FILE 'REGISTRY' ENTERED AT 00:23:30 ON 25 JUL 86
COPYRIGHT 1986 BY THE AMERICAN CHEMICAL SOCIETY

⇐ E CHLORHEXIDINE/CN ←————————————————————(2)
E1   1 CHLORHEXADOL/CN
E2   1 CHLORHEXAMIDE/CN
E3   1 CHLORHEXIDINE/CN
E4   1 CHLORHEXIDINE 3-HYDROXY-2-NAPHTHOATE/CN
E5   2 CHLORHEXIDINE ACETATE/CN
  *
  *
E12   1 CHLORHEXIDINE DIBENZOATE/CN

⇐ E 25 ←————————————————————————————(3)
E13   1 CHLORHEXIDINE DICAPROATE/CN
  *
  *
E37   1 CHLORHEXIDINE ISOPHTHALATE/CN

⇐ E 25 ←————————————————————————————(4)
E38   1 CHLORHEXIDINE LACTATE/CN
  *
  *
E62   1 CHLORIC ACID, ALUMINUM SALT/CN

⇐ S E3 OR E5 OR E9 OR E11 OR E19 OR E20 OR E24 OR E32 OR E33
OR E42
            1 CHLORHEXIDINE/CN ←————————————(5)
            2 "CHLORHEXIDINE ACETATE"/CN
            1 "CHLORHEXIDINE DI-D-GLUCONATE"/CN
            1 "CHLORHEXIDINE DIACETATE"/CN
            1 "CHLORHEXIDINE DIGLUCONATE"/CN
            1 "CHLORHEXIDINE DIHYDROCHLORIDE"/CN
            1 "CHLORHEXIDINE DILACTATE/CN
            1 "CHLORHEXIDINE GLUCONATE"/CN
            1 "CHLORHEXIDINE HYDROCHLORIDE"/CN
            1 "CHLORHEXIDINE TARTRATE"/CN
L3   7 CHLORHEXIDINE/CN OR "CHLORHEXIDINE
            ACETATE"/CN OR "CHLORHEX"

⇐ FILE CA

FILE 'CA' ENTERED AT 00:27:42 ON 25 JUL 86
COPYRIGHT 1986 BY THE AMERICAN CHEMICAL SOCIETY

⇐ S L2 STEPS ◀——————————————————————————(6)
L4 (403)L1
L5 (26218)LONG
L6 (2963)ACTING
L7 (2947)SUSTAINED
L8 (55498)RELEASE
L9 (153811)CONTROL?
L10 (55498)RELEASE
L11 (8566)SLOW
L12 (55498)RELEASE
L13 (2806)DELIVERY
L14   1 L1(L)(LONG(W)ACTING OR SUSTAINED(W)RELEASE
OR CONTROL?(W)RE

⇐ S L3(L)(L5(W)L6 OR L7(W)L8 OR L9(W)L10 OR L11(W)L12
OR L13)
            1111 L3
          26218 LONG
           2963 ACTING
           2947 SUSTAINED
          55498 RELEASE
         153811 CONTROL?
          55498 RELEASE
           8566 SLOW
          55498 RELEASE
           2806 DELIVERY
L15   5 L3(L)(L5(W)L6 OR L7(W)L8 OR L9(W)L10 OR
L11(W)L12 OR L13)

⇐ D L15 1-5
    *
    *
⇐ LOG Y
STN INTERNATIONAL LOGOFF AT 00:32:10 ON 25 JUL 86

(2)   The Expand command is used in order to examine the contents of an in-
        verted file. It is used here to verify the presence of the various chlorhexi-
        dine salts in the Chemical Name Index. The system automatically expands
        12 terms and assigns E numbers to them.
(3), (4) The Expand command is used in order to examine an additional 25 terms
        (a total of 250 terms can be listed).
(5)   The Search command is used to search for specific chlorhexidine salts by
        using the corresponding E numbers.

(6)   The Search L2 STEPS command is used to create separate answer files to each of the components of answer file L2.

7.2.   Chemical reactions

How and where can one obtain information on a specific chemical reaction? For example, where and how could a photochemist obtain information about the photochemical decarboxylation of phenoxyacetic acid to anisole? This problem can be summarized by Reaction 7.1.

$$\text{(7.1)}$$

The problem could be approached from two different viewpoints: defining the problem simply as a photochemical decarboxylation of phenoxyacetic acid or looking at it as the photochemical formation (or synthesis) of anisole from phenoxyacetic acid.

Let us compare the search results using the two different approaches to the problem. For simplicity we shall limit our search period to the years 1967–71 which is the period covered by the 8th Collective Indexes of the *CA*.

The best source to use for such a search is the *CA*; using the *CA* one should use the 8th Collective Formula Index and/or the 8th Collective Subject Index (if the search was to cover the period from 1972 onwards the Chemical Substance Index would have to be used, as in 1972, the beginning of the 9th Collective Index period, the Subject Index was divided into the Chemical Substance Index and to the General Subject Index).

Let us first use the 8th Collective Formula Index. The *CA* Formula Index links the molecular formulae of all chemical substances indexed in the Chemical Substance Index (prior to 1972 the Subject Index) to the corresponding *CA* abstract numbers for documents in which the substances are reported. The index is arranged according to the Hill System, compounds having the same formula being arranged alphabetically according to their *CA* Index Name.

In order to use the Formula Index we have first to calculate the molecular formula of phenoxyacetic acid, which is $C_8H_8O_3$. Under this entry there are over 110 different compounds in the 8th Collective Formula Index; none of them is phenoxyacetic acid. This should not surprise us as phenoxyacetic acid is not a *CA* Index Name, the *CA* Index Name for phenoxyacetic acid is acetic acid, phenoxy-. Indeed, one of the first compounds listed under $C_8H_8O_3$ is acetic acid, phenoxy-[122-59-8], followed by 48 entries. What should our next step be? Should we examine all 48 entries in order to determine their relevance to our problem, or should we abandon this search tool and use instead the corresponding Subject Index? It would take 3–4 hours to examine all 48 entries (an average of about 4 min per entry), but it would take only

148

a few minutes to obtain the desired information using the 8th Collective Subject Index.

Using the 8th Collective Subject Index we easily find the heading Acetic acid, phenoxy-[122-59-8] (the number in the square brackets is the compound CAS RN), followed by 48 different entries. These entries are composed of descriptive phrases (known as text modifications) and abstract numbers. All entries are listed alphabetically according to the descriptive phrases. There is one entry without a text modification—only an abstract number, which follows immediately the compound Registry Number. As was discussed earlier (p. 98) it implies that the synthesis of the substance is reported in the original reference. The text modification describes the way in which the compound (in our case acetic acid, phenoxy-) is mentioned in the primary source.

Our interest is in the photochemical decarboxylation of the compound, so we should look for relevant text modifications such as decarboxylation, photochemistry, or photolysis. Indeed some relevant entries were found examining the various text modifications appearing under acetic acid, phenoxy- (Fig. 7.1).

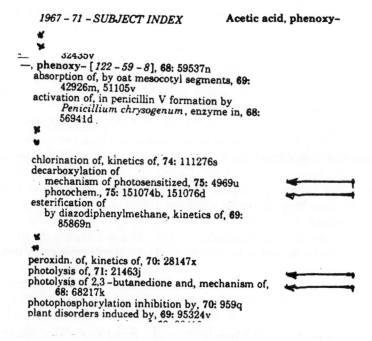

1967 - 71 - SUBJECT INDEX   **Acetic acid, phenoxy-**

—, phenoxy- [*122 - 59 - 8*], 68: 59537n
  absorption of, by oat mesocotyl segments, 69:
      42926m, 51105v
  activation of, in penicillin V formation by
      Penicillium chrysogenum, enzyme in, 68:
      56941d

  chlorination of, kinetics of, 74: 111276s
  decarboxylation of
    : mechanism of photosensitized, 75: 4969u
      photochem., 75: 151074b, 151076d
  esterification of
      by diazodiphenylmethane, kinetics of, 69:
        85869n

  peroxidn. of, kinetics of, 70: 28147x
  photolysis of, 71: 21463j
  photolysis of 2,3 -butanedione and, mechanism of,
      68: 68217k
  photophosphorylation inhibition by, 70: 959q
  plant disorders induced by, 69: 95324v

Figure 7.1   *CA* Subject Index

It was clear that the first three entries are relevant and the original references could be obtained by examining the abstracts 75:4969U; 151074b;

151076d, respectively [R.S. Davidson and P.R. Steiner, *J. Chem., Soc. C*, 1682 (1971); R.S. Davidson, K. Harrison and P.R. Steiner, *J. Chem. Soc. C*, 3480 (1971); R.S. Davidson and P.R. Steiner, *J. Chem. Soc. D*, 1115 (1971)]. On the other hand, the relevance of the last two entries was not clear and could be decided only after reading the original papers [D.P. Kelly, J.T. Pinhely, and R.D.G. Riby, *Aust. J. Chem.*, **22**, 977 (1969); E.J. Baum and R.O.C. Norman, *J. Chem. Soc. B*, 227 (1968)]; as it turned out those last two entries were not relevant to our topic.

The desired information was obtained within a few minutes using the Subject Index, whereas retrieval of the identical information using the Formula Index would have taken several hours. Doing a manual search one should preferably use the Chemical Substance Index (or the Subject Index prior to 1972) rather than the corresponding Formula Index. The Formula Index should be used to determine the *CA* Index Name of a compound whenever one has problems determining the 'correct' Index Name. However, if one uses the Formula Index and locates more than five or six entries to the compound, one should turn to the corresponding Chemical Substance Index in order to simplify the search.

The same search could be done online, using either a *CA* Search File corresponding to the 8th Collective Period (e.g. Dialog File 308, SDC File CAS 67) or the full *CA* File (e.g. Dialog File 399, STN *CA* File) while limiting the search to the 8th Collective Period. A search on Dialog File 308 is shown below.

File308:CA Search - 1967-1971
(Copr. 1986 by the Amer. Chem. Soc.)
   Set Items Description
   - - - - - - - - - - - - - - - - - -

? S RN = 122-59-8 AND DECARBOXYLATION AND (PHOTOCHEM? OR PHOTOSENSIT? OR PHOTOLYSIS)

   47 RN = 122-59-8
   711 DECARBOXYLATION
  16684 PHOTOCHEM?
  7289 PHOTOSENSIT?
  5015 PHOTOLYSIS
   1 3 RN = 122-59-8 AND DECARBOXYLATION AND (PHOTOCHEM? OR PHOTOSENSIT? OR PHOTOLYSIS)
? T1/1/1-3

1/1/1-3
75151076 75151074 75004969

Searching for the formation of anisole by photochemical decarboxylation of phenoxyacetic acid we can use as before either the 8th Collective Formula Index or the 8th Collective Subject Index. In order to use the Formula Index we have to calculate the formula of the anisole, which is C_7H_8O. Under C_7H_8O in the 8th Collective Formula Index one finds anisole (in the 9th Collective Index anisole is replaced by the *CA* Index Name, benzene, methoxy-) followed by its Registry Number [100-66-3] and a note, see Subjext Index. This means that there are a large number of entries to the compound (in our case, anisole). It will be useless to give such a large number of entries without any descriptive phrases. Therefore, the user is directed to use the corresponding Subject (or Chemical Substance) Index.

Using the 8th Collective Subject Index we were unable to find any relevant entry under anisole using descriptive phrases such as synthesis, preparation, formation, photochemical, or photochemistry. Furthermore, an online search results in obtaining no retrieval at all. Why? Did some error occur during the preparation of the Indexes, or was perhaps something wrong with the way the problem was presented? Presenting the problem as a photochemical formation, or synthesis, of anisole from phenoxyacetic acid means that we are interested in a photochemical synthesis which is competitive with other synthetic methods, giving good yield and a pure product. Is this the case? All of the entries found in the search of the photochemical decarboxylation of phenoxyacetic acid deal with the various products of the photochemical reactions of phenoxyacetic acid, one of the products being anisole. Its yield varied from less than 1% to about 80% depending on the reaction conditions and the sensitizer used. None of the above reactions (even the one with the high yield) could serve as a synthetic route to anisole. Here we are getting an answer to the question asked—during the period 1967-71 no photochemical synthetic route to anisole (via decarboxylation of phenoxyacetic acid or any other method) was described. However, during that period the photochemical decarboxylation of phenoxyacetic was studied (*CA*, 75:4969U; 151074b; 151076d) and anisole was found to be one of the reaction products.

Here we have seen the importance of the accurate definition of the problem.

A completely different reaction problem is the question, what is the Rydon–Smith reaction? What really is this reaction? Our organic chemist who needs the answer thinks that it has to do with detection and/or determination of peptide bond(s) in peptides and/or protein(s). It seems to him that maybe he should examine a book dealing with analysis and/or techniques in peptide and/or protein chemistry. Looking for such a book in his own institution's library he found the second revised and expanded edition of *Techniques in Protein Chemistry* by J. Legget-Bailey, published by Elsevier Publishing Company, Amsterdam, 1967. In the Subject Index of the book he found an entry 'peptid(s) detection, 31'. Indeed on page 31 he found the following: 'When sprayed with even high concentrations of ninhydrin peptides often show up very feebly. Cyclic peptides,... e.g. *N*-acetyl, *N*-methyl give very little or no colour with ninhydrin. In these cases exposure to chlorine gas to form the labile N-Cl compounds has been proposed by Rydon and Smith[63]. The reference to the original procedure (reference 63) is: H.N. Rydon and R.W.G. Smith, *Nature*, **169**, 922 (1952).

What would have happened if our chemist did not know that the above reaction had to do with the detection of peptide bond(s). He may be using the *CA*, but searching manually using the General Subject and the Subject Index gave negative results. Even an online search did not give any results, as can be seen:

File 399:CA SEARCH 1967-1986 UD = 10502
(Copr. 1986 by the Amer. Chem. Soc.)

  Set Items Description
  - - - - - - - - - - - - - - - - -
?S RYDON (W) SMITH
    1 RYDON
    611 SMITH
  S1 **0 RYDON(W)SMITH**
?LOGOFF
  23jul86 06:32:16 User016087

However our chemist decided to search the Abstracts Field which could be done only on the STN *CA* File. In doing so he obtained four relevant answers.

= FILE CA
FILE 'CA' ENTERED AT 07:25:43 ON 23 JUL 86
COPYRIGHT 1986 BY THE AMERICAN CHEMICAL SOCIETY

⇐ S (RYDON (W) SMITH)/AB
    4 RYDON/AB
    2203 SMITH/AB
L1   **4 (RYDON(W)SMITH)/AB**

⇐ D L1 1-4 CAN
CA98(9):67804g
CA90(23):182217q
CA87(3):20536d
CA84(1):3243b

In the various abstracts it was said: '...Its colour reaction was pos. for Rydon–Smith, Sakaguchi, and...'; '...gave pos. Rydon–Smith, Sakaguchi, and...'; 'It showed pos. tests with Sakaguchi, Dragandorff, and Rydon–Smith'; 'ninhydrin and Rydon–Smith reactions were pos...'. A reference to the Rydon–Smith reaction could be located in any of the four original articles [*Agric. Biol. Chem.*, **46**, 2979 (1982); **43**, 243 (1979); *J. Antibiot.*, **30**, 330 (1977); **28**, 828 (1975)].

One has to realize that, in searching for other name reactions or name tests, results could be different. The search for the Mcfadyen–Stevens reaction in Dialog File 399 yields 13 relevant answers.

File 399:CA SEARCH 1967-86 UD = 10502
(COPR. 1987 BY THE AMER. CHEM. SOC)

Set Items Description
- - - - - - - - - - - - - - - - - -
?S MCFADYEN(W) STEVENS
   13 MCFADYEN
   328 STEVENS
 S1   13 MCFADYEN(W)STEVENS

The same results were obtained searching the STN CA File in the basic index; however, three additional relevant answers were retrieved searching the abstracts field of the *CA* File on STN.

FILE 'CA' ENTERED AT 05:26:32 ON 23 JUL 87
COPYRIGHT 1987 BY THE AMERICAN CHEMICAL SOCIETY

⇐ S MCFADYEN(W)STEVENS
   13 MCFADYEN
   330 STEVENS
L1   13 MCFADYEN(W)STEVENS

⇐ S (MCFADYEN(W)STEVENS)/AB
   9 MCFADYEN/AB
   452 STEVENS/AB
L2   8 (MCFADYEN(W)STEVENS)/AB

⇐ S L2 NOT L1
L3   3 L2 NOT L1

The general attitude towards searching the *CA* File on STN should be that the abstracts field should be searched only if there are negative results when searching the basic field, or whenever an absolutely complete recall is required.

7.3. Concepts

An electrochemist is interested in obtaining information about heterogeneous redox oscillation reactions where although driven by an external source of electricity, no external power source is necessary. Our chemist knew of J. Keizer's work in which he called the phenomenon—beating mercury heart [*J. Am. Chem. Soc.* **101**, 5637 (1979)]. A manual search of the *CA* for beating mercury heart did not give anything (as a matter of fact, this was expected by our chemist). An online search results in three relevant answers, all of them Keizer's works [*J. Am. Chem. Soc.*, **101**, 5637 (1979); *Proc. Natl. Acad. Sci. USA*, **71**, 4477 (1974); S.W. Lin, Ph.D. Thesis—*Diss. Abs. Int. B*, **38**, 5965 (1977)—S.W. Lin was a graduate student of J. Keizer and coauthor of the preceding two papers]. Expansion of the search to mercury electrochemical oscillation(s) retrieved 12 answers of which only 5 were relevant (three of them obtained earlier searching for beating mercury heart). Further expansion to electrochemical oscillation(s) retrieved an additional 76 answers; none of them were relevant.

⇐ FILE CA
FILE 'CA' ENTERED AT 00:48:46 ON 25 JUL 86
COPYRIGHT 1986 BY THE AMERICAN CHEMICAL SOCIETY

⇐ **S BEATING(W)MERCURY(W)HEART**
776 BEATING
45688 MERCURY
68684 HEART
L1   3 BEATING(W)MERCURY(W)HEART

⇐ **S (MERCURY OR HG OR 7439-97-6)(L)ELECTROCHEM?(L)**
(OSCILLATOR# OR OSCILLATION#)
45688 MERCURY
3079 HG
30269 7439-97-6
66558 ELECTROCHEM?
9189 OSCILLATOR#
8478 OSCILLATION#
L2   12 (MERCURY OR HG OR
7439-97-6)(L)ELECTROCHEM?(L)(OSCILLATOR#

⇐ **S L2 NOT L1**
L3   9 L2 NOT L1

⇐ **S ELECTROCHEM?(L)(OSCILLATOR# OR OSCILLATION#**
66558 ELECTROCHEM?
9189 OSCILLATOR#
8478 OSCILLATION#
L4   88 ELECTROCHEM?(L)(OSCILLATOR#
OR OSCILLATION#)

154

⇐ S L4 NOT L2
L5   76 L4 NOT L2

Our chemist was sure that more relevant work on the subject was published during the last few years. In order to prove his point he decided to search the Science Citation Index (SCI) using Keizer's work [*J. Am. Chem. Soc.*, **101**, 5637 (1979)] as a starting point. He started with the 1985 Citation Index examining every year volume down to the year 1978. He found one author that cited Keizer's article in 1984, two in 1983, one in 1982, and one in 1980; altogether this specific article was cited five times since its publication. In order to get full information about those articles our chemist would have to turn to the articles themselves, locate them in the *CA*, and read their abstracts, or he may get some partial information by turning to the SCI Source Index where he could get a full bibliographic information (including the titles) of those five articles. Indeed those five papers are relevant to our electrochemist—two of them he obtained earlier searching the *CA*; the other three were new.

The same search could have been carried online, cutting drastically on search time and eliminating working with the small print used by SCI.

Let us consider a graduate student working towards his Ph.D. in physical organic chemistry who is interested in Möbius systems in the ground state. Our student was able to obtain from a colleague two relatively old references: E. Heilbronner's paper 'Huckel Molecular Orbitals of Möbius Type Conformations of Annulenes' [*Tetrahedron Lett.*, 1923 (1964)] and Zimmerman's review 'The Möbius-Huckel Concept in Organic Chemistry' (*Acc. Chem. Res.*, **4**, 272 (1971)]. He wanted to know what had been reported (experimentally or theoretically) on such systems since 1971.

The first step in the search is to use the *CA Index Guide* in order to locate any cross-references to Mobius, Moebius, or Möbius. When our student examined the *Index Guide* he was not able to locate any cross-references to the above terms. The second step is to use the General Subject Index of Volumes 102-96, the 10th and the 9th Collective General Subject Index, and the 8th Collective Subject Index (the search was carried out during January 1986, by that time the printed indexes of Volume 103 were not yet available) using the same search terms as before. Unfortunately, our chemist was not able to locate any relevant entry. This was unexpected as he already had one reference and he should have been able to locate it using the 8th Collective Subject Index (indeed, by using the corresponding Author Index he was able to locate the *CA* entry to Zimmerman's review— 75 : 129232e).

This situation made our student suspicious and he worried whether he was really using the accepted term(s) in his search. He turned to the latest version of *Glossary of Terms Used in Physical Organic Chemistry* compiled by V. Gold [*Pure Appl. Chem.*, **55**, 1281 (1983)]. There under the term Moebius Aromaticity, he found the following definition: 'A monocyclic array of orbitals on which there is a single out of phase overlap (or more generally, an odd number of out of phase overlaps) reveals the opposite pattern of aromatic character to Huckel systems: with 4n electrons it is stabilized (aromatic) whereas with $4n+2$ it is destabilized (antiaromatic). No examples of ground-state Moebius systems are known but the concept has been applied to transition states of pericyclic reactions. The name is derived from the topological analogy of such an arrangement of orbitals to a Moebius strip.' The definition was followed by two references—the same references which he had been given by his colleague.

Based on the above definitions, our student collected 11 search terms:

Antiaromatic system(s)
Antiaromaticity
Aromatic system(s)
Aromaticity
Charge distribution
Electron energy
Electron properties
Electron system(s)
Energy level distribution
Hueckel (or Huckel or Huckel)
Orbitals

Using the latest edition of the *Index Guide* he was able to locate cross-references to three of his search terms:

Huckel molecular orbitals method   — molecular orbitals

Huckel rule   — aromaticity, mathematics

Orbitals   — atomic orbitals, charge distribution, molecular orbitals

At this stage he decided to gather more information about his original search term (Möbius) in order to add, if needed, more search terms to his profile. He could not find any information in the latest edition of H. Bennett's *Concise Chemical and Technical Dictionary* (published by Chemical Publishing Company, New York) or the latest edition of G.G. Hawley's *Condensed Chemical Dictionary* (published by Van-Nostrand, Rheinhold, New York). This really did not surprise him as those dictionaries contain definitions of chemical terms, information about chemicals of technical importance, data on a large number of chemicals, and identification of a large number of commercial products. Turning to more general scientific sources our chemist examined the latest editions of Van-Nostrand's *Scientific Encyclopedia* and

McGraw-Hill's *Dictionary of Scientific and Technical Terms*. Although he could not find any information in the first source he was able to gather some data from the latter one (Mobius function and Mobius strip were the terms he found). Based on these data he decided to add one additional search term—Graph Theory. From the *Index Guide* he learned that graph theory is indexed under mathematics.

Our chemist had 13 search terms:

Antiaromatic system(s)
Antiaromaticity
Aromatic system(s)
Aromaticity
Atomic Orbitals
Charge distribution
Electron configuration
Electron energy
Electron properties
Electron system(s)
Energy level distribution
Mathematics
Molecular orbitals

that he was looking for in the General Subject Index of Volumes 102–96, the 10th and 9th Collective General Subject Index, and the 8th Collective Subject Index. He was able to locate 26 relevant entries to 21 references under the following terms:

Antiaromaticity	2 entries
Aromaticity	2 entries
Electron configuration	4 entries
Energy level distribution	3 entries
Mathematics	7 entries
Molecular orbitals	8 entries

At that point our graduate student had 20 new relevant references (one of the references he obtained was Zimmerman's review in the *Acc. Chem. Res.* which had been given to him by his colleague). He was also completely exhausted from the search, which had taken him about three days! He tried various search terms, some of them in many combinations, and could not face the *CA* Indexes any more. It is true that our young graduate student had not had any formal or informal training in searching the chemical literature in general, or in searching the *CA* in particular. However, even an experienced searcher would have spent many hours on this particular problem.

It is interesting to note that a simple online search would take just a few minutes and would result in the retrieval of 44 items.

File 399:CA SEARCH 1967-86
(Copr. 1986 by the Amer. Chem. Soc.)

```
   Set   Items    Description
   - - -  - - - - -  - - - - - - - - - -
?S MOBIUS   OR MOEBIUS
            4   MOBIUS
           42   MOEBIUS
   S1      44   MOBIUS OR MOEBIUS
?T1/3/1-44
```

1/3/1
   102113463 CA: 102(13)113463w DISSERTATION
A "strip-strategy" for the synthesis of molecular cylinders and Moebius
   *
   *
   *

JOURNAL: J. Amer. Chem. Soc. DATE: 1969 VOLUME: 91
NUMBER: 9 PAGES* 2330-8
CODEN: JACSAT LANGUAGE: English

?LOGOFF
```
          19feb86 00:27:03 User016087
      $9.27      0.103 Hrs File399
          $10.92 44 Types in Format 3
      $10.92 44 Types
      $20.19 Estimated cost this file
      $20.36 Estimated total session cost    0.110 Hrs.
Logoff: level 7.10.10 00:27:03
```

Examination of the 44 items shows that 28 were relevant ones (the other 16 dealt with the synthesis of Möbius strip molecules (5 items), natural and synthetic polymers (4 items), nuclear physics (2 items), chemiluminescence, liquid crystals, Mobius ring packing, phase transitions, and transition states in pericyclic reactions (1 item each)). Twenty-one items of the twenty-eight were retrieved by the manual search; thus the online search not only took much less time than the manual one, but it also retrieved 33% more relevant items. One should remember that the great difference (or the basic difference) between the manual and the online search is that whereas only controlled vocabulary terms can be used as search terms in the manual search, in the online search natural vocabulary terms as well as control vocabulary terms are used. This is of particular importance where one is searching in a new or emerging area (especially one in which many concepts are involved and/or not much work has been carried out), as it takes some time to evaluate appropriate control vocabulary terms.

People complain that information is not free any more—online searching costs money. However, it is money worth while spending. The online search of the Moebius problem on Dialog took 6.02 min., results in the typing of 44 references, and cost $20.19. This search would take an experienced chemist at least a whole day, costing his employer (in direct and indirect salary costs) much more than $20.19 (somewhat between $150 and $200 or maybe even more) and resulting in only partial retrieval of the desired information.

A completely different approach would be to use the *SCI* as a search tool. Our student could use either Heilbronner's original article [*Tetrahedron Lett.*, 1923 (1964)] or Zimmerman's review [*Acc. Chem. Res.*, **4**, 272 (1971)] as a starting point.

Heilbronner's paper was cited 78 times during the years 1971–85, but from the Citation Index itself one cannot know how many of the articles in which the work was cited are relevant. The problem with the Citation Index is that it gives only the first author's name followed by the journal reference. More information is found in the Source Index—a full bibliographic citation, the article's title, and the number of references cited in the article. However, in most cases one cannot decide about the relevance of the work; thus one has to turn to the original article or to the corresponding *CA* abstract. The search could be carried out online, thus saving time on the one hand and obtaining a bibliographic list of the cited papers on the other. The time saving in the above search is from 6–8 hours to 10–15 minutes. Out of the 78 citations, 14 were found to be relevant ones, 11 were located using the *CA*, while 3 were new ones that had not been retrieved before. Similar results were obtained starting with Zimmerman's review: out of the 239 citations only 11 were found to be relevant ones—nine were located using the *CA* while two were new ones (those two were also retrieved using Heilbronner's article as a starting point).

One may use the *SCI* Permuterm Index as a searching tool. Doing so one should remember that *SCI* is a multi-disciplinary tool, thus covering the various natural and life sciences. An online search of *SCI* for Mobius or Moebius results in about 180 items for the period 1974–85; out of those only 18 were relevant. All of those 18 items were retrieved using the *CA* as a search tool, 6 items retrieved in the *CA* were published in sources not covered by the *SCI*, and 2 items retrieved in the *CA* did not have the word Mobius or Moebius in their title (remember that bibliographic searching, online or manual, of the *SCI* is limited to the title only). The remaining 2 items out of the 28 retrieved in the *CA* were pre-1974 publications.

CHAPTER 8

Selecting a Bibliographic Data Base

In the various bibliographic searches we have been using up till now the *CA* data base is our first choice. Indeed *CA* is the dominating data base in chemistry and it is the main entry into sources of chemical information as well as chemistry-related information. *CA* covers not only chemistry and chemical engineering but also peripheral areas from subjects such as life sciences, material sciences, physics, etc. However, there are topics and subjects, mainly in multi-disciplinary areas, in which *CA* alone is not sufficient [e.g. it has been seen earlier (p. 142) that extra bibliographic information was obtained searching for 'island(s) of stability' in Physics Briefs and/or INSPEC in addition to *CA*].

How would one know which bibliographic data base or data bases to use in a search? Unfortunately, there is no answer to this question as it depends upon the search topic as well as upon the problem itself. One usually should use the *CA*; however, in the case of an interdisciplinary topic one or two other data bases covering the other discipline(s) should be used. The choice of the other data base(s) depends mainly upon the search problem.

Sometimes one should not even consider the *CA* as a first source. A biochemist seeking information about the various oxidases in peas' (*Pisum sativum*) roots is much better off searching *BIOSIS* than *CA*. *BIOSIS* data base contains 28 relevant references, *CA* data base contains 15, while only 9 references appear in the two data bases. Indeed, one should use both data bases in a search. Should one use a third data base as well?

In order to help the searcher solve this problem, while conducting his search online, some of the online vendors are producing a kind of information directory (e.g. *CROSS* on BRS, *DIALINDEX* on Dialog). These are file indexes of the various data bases which give the user access to the number of postings of a single term, a multi-word phrase, or a Boolean expression in preselected data bases. However, these directories give only peripherial coverage, giving information on which data base will yield the highest number of postings. One can use these directories to help in deciding which data base(s) should be selected for searching. The user has to keep in mind that the coverage period of each data base is different (so comparison should be made using yearly averages) and that he cannot test a long search profile nor develop and/or refine a search strategy on those directories.

An information scientist wanted to know what has been published in the year 1981 on the chemical compound dioxin (**8.1**) whose *CA* Index Name is dibenzo(b,e)(1,4)dioxin, 2,3,7,8-tetrachloro- [it is interesting to know that the synonym dioxin is used for another chemical compound as well—1,3-dioxan-4-ol, 2.6 dimethoxy-, acetate (**8.2**)].

8.1   **8.2**

Dioxin (**8.1**) is a chemical compound which is a trace contaminate in many products [e.g. in 2,4,5-T (**8.3**) which was widely used as a herbicide and was a major component of Agent Orange]. It is believed to be formed in minute quantities when anything containing carbon, hydrogen, oxygen, and chlorine is burnt, as well as a by-product during the hydrolysis of polychlorinated benzene. It is one of the most toxic materials known (its toxicity varies from animal to animal—LD_{50} of 1 nanogram per kilogram of body weight to guinea pigs; 5000 nanograms to hamsters), causing cancer in mice, cats, and probably in humans.

Few exposures of people to dioxin have been recorded, some of them before its toxicity and carcinogenicity were known. About 800 workers have been exposed to the compound in nine industrial accidents during the years 1948–76, unknown numbers of American soldiers have been exposed to Agent Orange in Viet-Nam in the years 1962–71, 37,000 people in Seveso and nearby communities were exposed when a reactor in a chemical factory making 2-4-5-trichlorophenol (**8.4**) went out of control in July 1976, contamination of dioxin was found in various places in the USA during the late 1970s-early 1980s (e.g. Times Beach Mo., Midland, Mich.).

8.3   **8.4**

The extreme toxicity of the compound drove scientists (environmentalists, toxicologists, physicians, chemists, biologists, etc.) and government agencies to deal with concentrations of the compound in the range of parts per billion or even lower concentrations.

Our information scientist decided to run his search online using 12 different data bases: one is chemistry— *CA*; one in the life sciences— *BIOSIS*; two in medicine— *EMBASE* and MEDLINE; two in agriculture— *Agricola* and *CAB Abstracts*; two in water sciences— *Aqualine* and *Aquatic Sciences and Fisheries Abstracts*; three in environmental sciences— *Enviroline, Environmental Bibliography*, and *Polution Abstracts*; as well as one interdisciplinary data base— *Scisearch* (the online version of *SCI*). The results of the search in the 12 different data bases are given in Table 8.1.

Table 8.1   Number of 1981 papers on dioxin cited by various data bases

	Number of citation	Percentage	Citation unique to the data base	Percentage of total coverage
CA	137	47.9%	55	19.2%
EMBASE	134	46.8%	46	16.1%
BIOSIS	97	33.9%	11	3.8%
Scisearch	95	33.2%	8	2.8%
MEDLINE	75	26.2%	10	3.5%
Pollution Abstracts	25	8.7%	7	2.4%
Environmental Bibliography	24	8.4%	2	0.7%
CAB Abstracts	21	7.3%	5	1.7%
Agricola	19	6.6%	—	—
Enviroline	13	4.5%	9	3.1%
Aqualine	8	2.8%	3	1.0%
Aquatic Science Abstracts	4	1.4%	—	—
Total	652		156	54.3%

Detailed examination of the 652 relevant citations which were retrieved showed that during the year 1981, 286 publications have appeared in the literature dealing with the various aspects of dioxin. The largest coverage of the subject was given by the *CA*, citing 47.9% of all the published material. Somewhat lower coverage is given by *EMBASE* (*Excerpta Medica*)— 46.8%. Seven data bases gave a coverage below 9%. It is interesting to know that 156 publications (54.3% of the total number) are reported in one data base only (on the average each work has been reported in 2.28 data bases). Two data bases— *Agricola* and *Aquatic Sciences and Fisheries Abstracts*—do not have any unique posting, eight other data bases give a very small number of unique postings (3.8–0.7% of the total number of relevant postings). Only two data bases have a large percentage of unique postings— *CA* (55 out of 286— 19.2%) and *EMBASE* (46 out of 286— 16.1%).

Not only does the *CA* have the highest number of unique postings and the best subject coverage but seven out of the other eleven data bases overlap the *CA* by more than 50%, and three of them (*Agricola, Environmental*

162

Bibliography, and *Aquatic Sciences and Fisheries Abstracts*) have an overlap
of 75% and more (Table 8.2).

Table 8.2   Overlaps of the 1981 dioxin citation of the various data bases with *CA*

	Number of citation	Overlap	Percentage
EMBASE	134	57	42.5%
BIOSIS	97	58	59.8%
Scisearch	95	48	50.5%
MEDLINE	75	39	52.0%
Pollution Abstracts	25	11	44.0%
Environmental Bibliography	24	18	75.0%
CAB Abstracts	21	15	71.4%
Agricola	19	17	89.5%
Enviroline	13	3	23.1%
Aqualine	8	3	37.5%
Aquatic Science Abstracts	4	3	75.0%

In order to obtain best results one should carry a multi data base search. In
fact while *CA* yields a 47.9% coverage and *EMBASE* yields 46.8% coverage
a combination of the two data bases results in a 75% coverage. In order to
increase the coverage one should add at least one additional data base. Which
one should it be? Another medical one *MEDLINE*? A biological one (*BIO-
SIS*)? An environmental one (*Pollution Abstracts* or maybe *Environmental
Bibliography*)? A multi-disciplinary one (*Scisearch*)? Intuition will tell us
that best additional coverage would be obtained using a multi-disciplinary
data base. Indeed, using *Scisearch* as a third data base gave 87.4% coverage,
while adding *MEDLINE* instead gave 82.5% coverage, *BIOSIS* — 82.2%, *Pollution Abstracts* — 80%, and *Environmental Bibliography* gave 78% coverage
only.

Out of the 286 publications 7 are patents while the others are articles
published in journals and proceedings. The 279 articles have been published
in 154 journals and proceedings, 140 of those journals (91%) being covered
by the *CA*. One has to keep in mind that only a very small number of all
sources covered by the *CA* are covered cover to cover. (This is true for
all data bases with very few exceptions— *Scisearch* or *SCI* is one of the few
exceptions.) Indeed, out of the eight articles on dioxin which were published
in *Science* during 1981, only three were reported in the *CA*; of seven articles
published in *Nature* during that period only two were reported in the *CA*;
Lancet and *Pharmacologist* are two of the many journals covered by the
CA, yet none of the relevant articles published in these journals (seven and
six, respectively) were included in the *CA*.

One may divide the dioxin literature into seven topics—analysis, safety,
toxicology, environmental aspects, biological aspects, medicine, and miscellaneous. The coverage of the first five topics by the *CA* is 44–69% (see Table
8.3) while the last two topics are not covered at all by the *CA*, something

which should not surprise us. The topic with the best coverage by the *CA* is of pure chemical aspects—analysis of dioxin (69% coverage) while topics which are chemically oriented—safety, toxicology—have a somewhat lower coverage by the *CA* (55% each). One has to remember that CAS Indexes and abstracts articles which possess some novelty in chemistry, thus usage of known analytical methods for detection and/or determination of dioxin would not be abstracted by the CAS, but would be reported by data bases such as *Environmental Bibliography*, *Pollution Abstracts*, etc. The picture changes somewhat when one examines the combined coverage of *CA* and *EMBASE*—92% coverage of the biological aspects, 82% of toxicology, 72% of analysis, and 71% of safety.

Table 8.3   1981 dioxin subject coverage by *CA* and *CA/EMBASE*

	Total	Chemical Abstracts		Chemical Abstracts and Exceptra Medica	
		Number	Percentage	Number	Percentage
Analysis	36	25	69%	26	72%
Safety	49	27	55%	35	71%
Toxicology	27	15	55%	22	82%
Environmental aspects	64	29	45%	38	59%
Biological aspects	92	41	44%	85	92%
Medicine	15	—	—	7	47%
Others	3	—	—	1	33%
	286	137	48%	214	75%

Multi-search gives us much better results in searching specific topics connected with dioxin, as can be seen in Table 8.4.

Five searches were carried out:

A. Formation of chemical compounds without dioxin contamination
B. Decomposition and destruction of dioxin
C. The Seveso incident
D. Formation of dioxin in incinerators
E. Dioxin contamination of water

The coverage by *CA* of these five topics was 75%, 65%, 41%, 66%, 43%, respectively. Unique citation by the *CA* to those topics was 75%, 50%, 33%, 33%, and 0%, respectively. Best coverage by two data bases is given by *CA* and *Scisearch* to topic A—100% coverage; by *CA* and *EMBASE* to topics B, C, and D—75% (15 out of 20), 58% (7 out of 12) and 89% (8 out of 9), respectively; and by *Aqualine* and *Environmental Bibliography* to topic E—86% coverage (6 out of 7).

Our information scientist decided that he would like to get a second data base to search for patents related to dioxin. Searching *Derwent World Patent*

Table 8.4   Coverage of specific topics related to dioxin in various data bases
(topics are listed in the text)

	A	B	C	D	E
CA	3(3)	13(10)	5(4)	6(3)	3
EMBASE	—	4(2)	3(2)	5(2)	2
BIOSIS	—	2	3	2	2
Scisearch	1(1)	3(1)	2	3	3
MEDLINE	—	3	3(1)	1	1
CAB Abstracts	—	3(2)	—	1	—
Pollution Abstracts	—	1	1(1)	1	2
Enviroline	—	1(1)	—	3(1)	1(1)
Environmental Bibliography	—	—	—	—	4(1)
Agricola	—	—	1	—	1
Aqualine	—	—	1(1)	—	3(2)
Aquatic Science Abstracts	—	—	—	—	2
Total	4	20	12	9	7

Index he found eight patents which were granted in 1981; seven of them were also retrieved from the *CA*.

Those results show clearly that when conducting a bibliographic search on a topic which is on the borderline of chemistry or chemical engineering or is multidisciplinary in nature one should use at least two data bases for search—the *CA* and at least one other data base relevant to the search topic. *CA* is the dominant data base in chemistry. In many other areas of science one cannot define a dominant data base; in physics we have among others *Physics Briefs* and *INSPECT*, in medicine *MEDLINE* and *EMBASE*, in agriculture CAB Abstracts and Agricola, etc. Which data base should one use as a supplementary one for a biochemical search—*BIOSIS* or maybe one of the medical ones (and if so which one—*MEDLINE* or *EMBASE*)? Maybe one should not use *CA* and instead use *BIOSIS* and one of the medical data bases. In order to be able to make the best choice, one should be familiar with the structure and coverage policy of the candidate data bases. We have seen the difference in retrieval results between the two agricultural data bases as well as between the two medical ones in the dioxin search; only coverage policies could be responsibile for such differences. *Agricola* and *CAB Abstracts* gave a similar number of relevant postings (19 and 21, respectively). However, only four of them are overlapping. Furthermore, no unique postings are retrieved in *Agricola* in comparison with five such postings in *CAB Abstracts*. A similar picture is found in comparing *EMBASE* and *MEDLINE*—134 relevant postings against 75, 46 unique postings (34%) in comparison to 10 (13%), overlapping of 53 postings, having 7 and 10 postings, respectively, on clinical medicine out of which only two are overlapping.

One should remember that while it is easy to obtain over 90% recall in a well defined chemical search topic using the *CA* the situation is different in the case of a multi-disciplinary topic or a topic on the border of chemistry and some other discipline. *CA* will usually yield 40–70% recall while a right selection of a second data base would increase our recall to 75–90%; getting the remaining 10–25% is usually a hard and tedious job.

CHAPTER 9

Synthetic Reaction Search

When one is interested in synthetic chemistry one's interests are in locating starting materials or intermediates for use in one's own synthetic programme (however a chemist might be interested in those compounds also as models for biological, physical, or chemical studies) and should find out whether the desired compound is commercially available. The quickest and usually the cheapest way to obtain a chemical compound is to buy it from one of the supply houses or from one of its manufacturers. If it has been found that the desired compound is not commercially available or too expensive, one should turn to the bench and synthesize it. In order to do so, one has to search to find out whether the synthesis of the compound has been described in the literature in order to use the described procedure for the synthesis. In case no satisfactory synthesis (or no synthesis at all) is described one should design the synthetic route to be carried out.

9.1. Locating sources of fine chemicals

One should turn first to the major suppliers, who hold thousands (some-times over 20,000) of fine chemicals in stock. Usually one would have the catalogues of such suppliers on one's desk, and thus prices and quality can be compared. This is a worthwhile policy as there are many cases in which differences exist in prices and/or quality. Whenever one does not have the various catalogues one should ask the suppliers for complimentary copies. These catalogues are also held by the purchasing departments of the various institutions. However, one has to remember not only that obtaining chemicals from abroad takes longer (unless they are stocked locally in a warehouse) but that shipping charges and duty must be added to the catalogue price.

In many cases the required compounds cannot be found in any of the major manufacturers' catalogues as they are really special or 'exotic' fine chemicals which are produced or distributed by small, speciality companies. However, many scientists are interested in such chemicals and would prefer to buy them instead of making them. Are there any tools that will enable one to obtain information about the commercial availability of all fine chemicals? There is an annual publication entitled *Chem Sources*, as well as *The Fine Chemicals Directory* (*FCD*) and *Chem Buy Direct*. When one fails to locate a commercial source for a chemical using the few catalogues one has available, one should consult the latest available edition of *Chem Sources* or *FCD*.

Chem Buy Direct which was published by Walter de Gruyer was an excellent directory at its publication time (as a matter of fact it was the best one in the years 1975–76). However, this directory was never updated nor published in a new revised edition and a very good part of its contents are valueless at present.

Chem Sources is published by the Directories Publishing Company, Clemson, SC, USA, in two parts: *Chem Sources—USA*, and *Chem Sources—Europe*. It includes over 1000 producers and distributors of fine chemicals in the USA and Europe as sources for about 110,000-120,000 chemicals. It also supplies information about the various producers and distributors (addresses, telephone numbers, etc.). An online version of *Chem Sources—USA* is available online on STN International as two Files—*CSCHEM* and *CSCORP*. *CSCHEM* contains information about the various chemical compounds available and their suppliers. *CSCORP* contains information about the various US companies (producers and distributors) that supply chemicals.

The *FCD* is published by Fraser Williams Scientific Systems. It covers over 100,000 chemicals produced by 48 producers. It supplies information about the various products as well as about the producers. An online version of FCD was available on Pergamon Infoline.

Let us consider an American physical chemist who needs galium fluoride trihydrate ($GaF_3.3H_2O$) for his research—from whom should he order the desired compound? First he has to find out if the compound is commercially available. This is done by examining catalogues of some of the major fine chemical producers and/or distributors (e.g. Aldrich, Eastman–Kodak, Fluka). He checked the latest available catalogue of Aldrich (1985/6) and found out that the desired compound (catalogue number 20-332-7) is available in 99.999% purity at a price of $46.25 for 5 g.

Consider an Israeli organic chemist who needs small amounts of *trans*-2-methyl-1,3-pentadiene (**9.1**) and 4-methyl-1,3-pentadiene (**9.2**). He checked the latest Aldrich catalogue to find out that while *trans*-2-methyl-1,3-pentadiene is available for sale (compound 11-110-4, 1 g sells for $15.30 and 5 g for $51.01) the 4-methyl-1,3-pentadiene is unavailable. His next step was to use the latest Fluka catalogue (1986/7). There he found that both compounds are sold by Fluka; compound **9.1** sells for SF75 for 5 ml (catalogue number 68277) and compound **9.2** sells for SF125 for 5 ml (catalogue number 68290).

9.1   **9.2**

Let us consider an organic chemist who is in need of a large amount (300–500 g) of Z,Z-2,4-hexadienedinitrile (**9.3**) as a starting material for the total synthesis of a natural product. He was not able to locate a producer of this compound by using various catalogues which he had. He decided to run an online search in the *FCD*; unfortunately compound **9.3** was not reported in the *Directory*.

Chemical compounds could be searched in the *FCD* by molecular formula, chemical name, and Wiswasser Line Formula Notation (WLN) (see p. 202). An example of a search for manufacture(s) of 5-fluoroindole (**9.4**) is given below:

9.3

9.4

FINE CHEMICALS DIRECTORY - VERSION 2
Copyright 1983, Fraser-Williams (Scientific Systems) Ltd
- Updated October 1985

There are now **50** catalogues from **48** suppliers
For full details of catalogues type: **?UPDATE**
Enter your request
/ S WL= T56 BMJ GF ←————————————————————(1)

SET   1: 1   WL= T56 BMJ GF ←————————————(2)
/ D 1F9 ←——————————————————————————————(3)

Item 1 ←——————————————————————————————(4)
Accession No: **25889**
Mol. Formula: **C8H6FN**

WLN & Suffix: **T56 BNJ GF**

Supplier	Cat.No.	Compound Name
ALDRICH	F00910-8	5-FLUOROINDOLE, 98%
COLUMBIA	F-1060	5-FLUOROINDOLE
FAIRFLD	F-168000	5-FLUOROINDOLE 99%
FLROCHEM	F03758	5-FLUOROINDOLE 97%
K & K	K 6152	5-FLUOROINDOLE
KOCHLT	2726	5-FLUOROINDOLE PURE
P & B	F03220	5-FLUOROINDOLE
SIGMA	F1751	5-FLUOROINDOLE

(1) The Search command is used to search for 5-fluoroindole in the WLN field (T56 BNJ GF is the notation of compound **9.4**).
(2) There is one entry in the data base.
(3) A Display command is used to display the full record (formate 9).
(4) The full record contains the *FCD* accession number, the molecular formula, the WLN, and a table listing suppliers, catalogue numbers, and information about the compound (names, salts, purity, etc.).

The online version of *FCD* has been replaced recently by *Chem-Quest* which is available on Orbit. This data base, which is based on the *FCD*, includes new fields such as CAS RN and prices. Thus searching could be carried out by RNs and prices of the various producers could be compared. Another feature of Chem-Quest is the electronic ordering of the desired compound(s) at the end of the search.

Our organic chemist realized that compound **9.3** is not commercially available and he would have to synthesize it. The first thing he has to do in such a case is to conduct a search for a procedure for the synthesis of the compound.

9.2. Procedures for the synthesis of chemical compounds

9.2.1. Organic compounds

The first source one should use when looking for a known synthetic procedure of an organic compound is the *Beilstein*. Our organic chemist knows that **9.3** is a functional derivative and its 'hydrolysis' will give the corresponding Registry Compound Z,Z-1,3-butadiene-1,4-dicarboxylic acid (**9.3**). Acyclic carboxylic acids are discussed in Volume 2. Examining the Cumulative Formula Index for Volumes 2 and 3 he found under $C_6H_4N_2$ an entry to hexa-2,4-dienedinitril-2 IV 2302. Indeed, in Volume 2 of the Fourth Supplement our chemist found all available information on compound **9.5**, at least up to the end of 1959. Among other things he found two syntheses of the desired compound, one starting with 3-chlorohexa-3(*trans?*)-enediamid (**9.6**) [*Justus Leibigs, Ann. Chem.*, **627**, 1 (1959)] and the other starting with *cis, cis*-muconatenediamide (**9.7**) [*J. Chem. Soc. C*, 385 (1966)].

9.5

9.6

9.7

After reading the original articles our chemist decided to continue his search as those synthetic procedures did not appeal to him—the starting materials were hard to get, the yields were low, and he preferred to avoid working with large amounts of sodium cyanide or phosgene. He decided to turn to the *CA* for a literature search for the period 1985–60. Using the *CA* Chemical Substance Index from 1985 backwards, he found a few entries to the synthesis of **9.3**. He decided to stop the search once he found an entry in the 10th Collective Index 88:136115r—'Cuprous Ion Catalysed Oxidative Cleavage of Aromatic *o*-Amines by Oxygen' [J. Tsuji and H. Takayanagi, *Org. Synth.*, **57**, 33 (1977)].

Turning to the original work he found a simple reaction starting with a readily available starting material, *o*-phenylenediamine, and a simple procedure, bubbling oxygen through a solution of copper(I) chloride in pyridine followed by addition of the *o*-phenylenediamine (1,2-diaminobenzene—**6,15**) and extraction of the product from the reaction mixture with methylene dichloride. The Cu$^+$-catalysed oxidative cleavage of *o*-phenylenediamine by gaseous oxygen gave **9.3** in 88–93% yield (Reaction 9.1).

$$ \text{(9.1)} $$

9.2.1.1. Organic Syntheses

Our chemist decided to use the *Organic Syntheses* procedure even before reading the original work. This is the usual attitude of organic chemists towards *Organic Syntheses*, as this serial publication is known for its high selectivity and standards. It is an annual compilation of syntheses of organic compounds, its emphasis in recent years being directed mainly to model compounds and procedures which illustrate important reaction types. Each reported procedure is checked and repeated by independent organic chemists (usually two, but sometimes more). Starting with Volume 50 a list of all the unchecked procedures received for publication during the year is inserted at the end of the volume. Since Volume 55 the list included only procedures which had been accepted for checking. These procedures are available from the secretarial offices of Organic Syntheses Inc. for a nominal fee. However, one has to remember that these procedures are unchecked, unedited, and will not necessarily be published. Starting with Volume 62 changes in presentation and distribution took place in order to shorten the times between submission and appearance of the accepted procedures. The printing is a direct reproduction of the manuscript; soft cover copies (without indexes) are distributed free of charge to the members of the Organic Chemistry Division of the ACS for their personal use.

In addition to the annual publication there are Collective Volumes (10 years collective volumes for Volumes 1–59 and five years collective volumes starting with Volume 60), which revise and update the annual volumes whenever necessary. It also contains a cumulative index to the ten- or five-year period (indexes are to Preparation According to Type of Reaction, Compounds, Formulae, Preparation and Purification of Solvents and Reagents, Apparatus, Authors, and General Index). Whenever Collective Volumes are available they should be used in preference to the annual volumes.

Many organic chemists start their search by using Organic Syntheses, consulting first its Collective Indexes. This Collective Indexes volume covers the first five Collective Volumes of *Organic Syntheses* and provide an entry to all compounds whose syntheses have been reported by original name of the compound used in *Organic Syntheses* and also by the latest *CA* Index Name. Other indexes are Reaction Type, Compound Type, Formula, Solvents, Reagents, Apparatus, Authors, and a General Index to all compounds.

9.2.2. Inorganic compounds

Let us consider a second-year chemistry student taking a course in chemical information. One of the questions he was asked in his library examination was to describe the best available synthesis of anhydrous magnesium chloride ($MgCl_2$). The problem looked a very simple one, as it seems to him that the answer could be found in *Gmelin*. Using the *Gmelin* Formula Index our student found under Cl_2Mg, preparation -27(Mg)Hb/B1 104-5. Thus the available information could be found in Volume 27/B1 of the main series on pages 104–105. Turning to Volume 27/B of *Gmelin* our student found a list of several synthetic methods: reaction of magnesium with chlorine, hydrochloric acid, or various chlorides (e.g. S_2Cl_2, PCl_3, PCl_5); reaction of magnesium oxide with chlorine, hydrochloric acid, thionyl chloride ($SOCl_2$), sulphuryl chloride (SO_2Cl_2, or phosgene ($COCl_2$); dehydration of the hexahydrate salt ($MgCl_2.6H_2O$); and decomposition and dehydration of the double salt $NH_4Cl.MgCl_2.6H_2O$. Volume 27/B of *Gmelin* was published originally in 1939 and Section B/1 has a coverage period of up to the end of 1937. As no supplements to Volume 27 have been published, our student had to look for another source. With the aid of the librarian he decided to use *Methodicum Chemicum*.

Methodicum Chemicum is a comprehensive critical review of chemical methods applied in scientific research. It is aimed mainly at the chemist, but also at scientists working on the borders of chemistry and the life sciences. There is emphasis on newer methods which have not been described adequately in other sources, while older proven methods and techniques are to be found in the discussion. The publication of the series started in 1973 and it is composed of three parts: *General Part*, *Systematic Part*, and *Special Part*. Original plans were for publication of 11 volumes; however, only five volumes have been published till 1978, no other volumes have been pub-

172

lished since then. In the *General Part* only Volume 1—*Analytical Methods*, was published in two parts: Part A—Purification, Wet Processes, Determination of Structure; Part B—Micromethods, Biological Methods, Quality Control, Automatization. In the *Systematic Part* three volumes were published: Volume 6,—*C-N compounds*; Volume 7, *Main Group Elements and Their Compounds*, Part A—Group 0 to IV Elements and Their Compounds, Part B—Group VA to VIIIA Elements and Their Compounds; Volume 8, *Preparation of Transition Metal Derivatives*. In the *Special Part* only Volume 11, *Natural Compounds*, was published in three parts: Nucleic Acids, Proteins, and Carbohydrates; Antibiotics, Vitamins, and Hormones; Steroids, Terpenes, and Alkaloids. Volume 6 described the synthesis of compounds with functional groups having carbon–nitrogen bonds, arranged according to the resultant substance class; Volumes 7 and 8 describe the isolation and purification of the main group elements and the transition metals, and syntheses of their inorganic and organic compounds. All the published volumes have been published in two editions: German and English.

Our student turned to Volume 7/A of *Methodicum Chemicum—Main Group Elements and Their Compounds*—Group 0 to IV Elements and Their Compounds. On examining the Table of Contents of the English edition he found that Chapter 7.3 is devoted to Magnesium Halides (page 86). Indeed on page 86 he found a paragraph describing various synthetic methods for the desired compound followed by the corresponding references: reaction of magnesium with 1,2-dichloroethane or mercury chloride ($HgCl_2$); reaction of hydrochloric acid, chlorine, or benzyl chloride with certain Grignard compounds; and decomposition and dehydration of $NH_4Cl.MgCl_2.6H_2O$ in the presence of hydrochloric acid. The last method, although already found before in *Gmelin* seems to be the most promising synthesis as its source was *Inorganic Syntheses* [*Inorg. Synth.*, **6**,9 (1960)].

9.2.2.1. Inorganic Syntheses

All that has been said about *Organic Syntheses* could also be said about *Inorganic Syntheses*. It is an annual publication of checked procedures for the synthesis of important inorganic compounds, covering various broad, active areas of inorganic chemistry. One limitation of the series is that there are no Collective Volumes or Collective Indexes to the 26 volumes published so far. In order to search *Inorganic Syntheses* directly for the synthesis of a specific compound one has to examine the Indexes or the Tables of Contents of the whole series volume by volume.

Our student wanted to bring his search up to date. This could be done easily by covering the period from 1972 onward either by using the *CA* Indexes or the *SCI* (remember that literature coverage of *Methodicum Chemicum* Volume 7/A extends only up to the end of 1971).

In using the *SCI* for such a search our student is looking for who cited D. Bryce–Smith, *Inorg. Synth.*, **6**,9 (1960). He assumed that two types of works would cite the *Inorganic Syntheses* procedure—those that made $MgCl_2$ by a new or modified route, thus listing known existing methods for the synthesis of the compound, and those that synthesized the compound according to the Bryce-Smith procedure. (Indeed, the above synthesis has been cited three times. All of the citations referred to the synthetic procedure of the preparation of $MgCl_2$, none of the three give a new synthesis or a modification of an old synthesis of $MgCl_2$.)

9.2.3. Searching the *CA* for synthetic procedure

A manual search of the *CA* for a synthetic procedure for the preparation of a chemical compound is carried out using the Chemical Substance Index (prior to 1971 the Subject Index is used). It has been mentioned earlier that whenever a *CA* Index Name is not followed by a descriptor it means that the emphasis in the source article is on the synthesis of the compound. However, one should also look for descriptors dealing with synthesis (or their synonyms), e.g. synthesis, preparation(s), formation, manufacture(d), manufacturing. An online synthetic search should be carried out using a profile search having the chemical compound CAS RN and all the synonyms for synthesis, combined with those two search terms with a proximity operator that limits their combination to the same data field (F in Dialog, L in STN, etc.). However, in order to simplify synthetic searches various vendors developed unique algorithms for such searches. Indeed, using Dialog or STN one can retrieve synthetic procedures just by introducing the CAS RN with a P as a suffix. Using this approach one has to remember two things: (a) this method retrieves most synthetic procedures but not necessarily all of them (although in most cases the retrieval in 100%); (b) each vendor has developed its own algorithm and there are differences in the algorithms among the various vendors, as could be seen in the search for locating synthetic procedures for the synthesis of chenodeoxycholic acid (**9.8**)—the compound CAS RN is 474-25-9.

$CHMeCH_2CH_2CO_2H$

9.8

?B399
24feb86 05:15:53 User016087
$0.17 0.007 Hrs Filei
$0.17 Estimated total session cost 0.007 Hrs.

File 399:CA SEARCH 1967–1986 UD = 10406
(Copr. 1986 by the Amer. Chem. Soc.)

174

Set Items Description
- - - - - - - - - - - - - - - - - - - -
?S RN = 474-25-9P
    S1   30 RN = 474-25-9P

⇐ FILE CA
FILE 'CA' ENTERED AT 05:19:32 ON 24 FEB 86
COPYRIGHT 1986 BY THE AMERICAN CHEMICAL SOCIETY

⇐ S 474-25-9P
L1   64 474-25-9P

An examination of the results of the search shows that while 4 out of the 30 results obtained using Dialog File 399 are unique to this file, 38 out of the 64 results obtained using the STN CA File are unique to STN. A close examination of the results shows the difference between the two algorithms: 35 out of the 38 unique answers to STN have a text modification to the CAS RN (474-25-9) starting with 'formation of...', 'isolation of...', 'purifin. of...', 'total (or partial) synthesis of...'. In such cases the Dialog algorithm does not add the suffix P. The other three cases are completely irrelevant, the text modification starts with 'liver (or colon) carcinogenesis from...', 'teratogenesis in relation to...'. As it turned out, for some reason the STN algorithm adds the suffix P to a CAS RN if an ending genesis in a text modification of an RN exists when the abstract appears in some specific sections of the *CA*. As for the unique Dialog results, in three of them the word prepn. appears at the end of the text modification and not at the beginning—the text modification starts with 'dtem. of...'; thus those are analytical works and not synthetic ones (as a matter of fact the 'prepn.' refers to the preparation of columns for separations or for the preparation of conjugates for immunoassays).

When one includes formation, isolation, purification, and/or total synthesis as a preparative descriptor one should use the STN algorithm; in the case of using the Dialog algorithm one should add the missing search terms. Conversely, when one is not interested in formation, isolation, or purification, one may be better off using the Dialog algorithm (however, one should be aware that one would be missing total or partial synthesis of the compound if those terms are used as text modifications).

How could an industrial chemist find information about the production of benzoic acid by carboxylation of benzene (Reaction 9.2). Let us limit our search to the years 1972-77 (the period covered by the 9th Collective Indexes of the *CA*).

One can look at the problem as the reaction of benzene with carbon dioxide (i.e. carboxylation of benzene) to give benzoic acid. On the other hand one can look upon the problem as our industrial chemist sees it: the

$$\text{benzene} + CO_2 \longrightarrow \text{benzoic acid} \qquad (9.2)$$

production of benzoic acid by carboxylation of benzene—the synthesis of benzoic acid from benzene and carbon dioxide.

The chemist should start his search using the 9th Collective Chemical Substance Index. Many changes in the *CA* indexing system took place at the beginning of the 9th Collective Period (at the beginning of 1972). Some of these changes have already been discussed (e.g. the splitting of the Subject Index, nomenclature changes); another change is the subdivision of chemical substances with large numbers of entries. In the case of a compound with many entries, seven subgroup headings are used in order to group together broad subjects of related interest. The headings are Analysis (for methodology of detection and/or determination of the substance), Biological Studies (for formation, occurrence, properties, and processes of the substance in biological systems and for its biochemical uses), Preparation (synthesis, manufacture, formation, and purification excluding biosynthesis), Properties (chemical and physical properties), Reactions (chemical changes which result in products differing chemically from the starting material), and Uses and Miscellaneous (applications, industrial usages and processes, as well as studies not otherwise classified).

When searching for the reaction of benzene with carbon dioxide in the Chemical Substance Index one should use the subgroup or subheading Reactions under the heading (the substance name) Benzene. The appropriate index text modifications are carbon dioxide and/or carboxylation. Although no entries could be found using the term carbon dioxide, three entries were located under the term carboxylation:

benzoic acid from, 77:P164245z

catalysis for, 81:P25389v

in presence of PdCl$_2$ and CH$_3$COONa, benzoic acid and biphynyl from, 84:P135245p

All three entries appear to be of interest and indeed, after reading the abstracts, our industrial chemist ordered the original patent documents.

Another approach is to look under Benzoic Acid, subheading Preparation, using benzene, carbon dioxide, carboxylation, and/or manufacturing as index text modifications. Although no entries could be found using carbon dioxide as a text modification, each of the other three text modifications retrieves one entry:

from benzene, catalysis for   81:P25389v

by carboxylation of benzene, catalysis for   84:P135245p

manuf. of
    by carboxylation of benzene   77:P164245z

Those three entries are the same as those which were located using the reaction of benzene with carbon dioxide approach.

The same search could be carried out online, searching, for example, Dialog File 309 (which is corresponding to the 9th Collective Index Period).

File309:CA Search - 1972-1976
(Copr. 1986 by the Amer. Chem. Soc.)
          Set Items Description
          - - - - - - - - - - - - - - - - - -
? S RN = 65-85-0P AND RN = 71-43-2

          171 RN = 65-85-0P
         7985 RN = 71-43-2
     1   13 RN = 65-85-0P AND RN = 71-43-2
? S CARBON(W)DIOXIDE OR CARBONDIOXIDE OR
CARBOXYLATION

          10202 CARBON(W)DIOXIDE
            18 CARBONDIOXIDE
           756 CARBOXYLATION
     2 10865 CARBON(W)DIOXIDE OR CARBONDIOXIDE OR
              CARBOXYLATION
? C 1 AND 2

     3   5 1 AND 2
? T3/1/1-5

3/1/1-5
84135245 81025389 81005175 78015725 77164245

The online search results in five 'relevant' answers: three of them have been retrieved manually and two are new ones. Close examination of those two additional entries shows that they are really flash positive answers. One (78:15725) deals with thiocarboxylations of aromatic compounds; it has been retrieved because it deals with the preparation of benzoic acid (65-85-OP), reaction(s) of benzene (71-43-2), and one of its index entries is: carboxylation, thio of benzene dervs. The other one (81:5175) deals with the thermal decomposition of benzoyl peroxide which results in the formation of benzoic acid (65-85-OP), benzene (71-43-2), and carbon dioxide.

One should also remember that there are cases in which too many search terms in an online search result in a low (sometimes very low) recall. Such was the case with a biotechnologist who was interested in the enzymatic synthesis of L-aspartic acid from fumaric acid by aspartase.

⇐ FILE CA
FILE 'CA' ENTERED AT 00:44:51 ON 03 MAR 86
COPYRIGHT 1986 BY THE AMERICAN CHEMICAL SOCIETY

⇐ S 56-84-8P AND 110-17-8 ←————————————————————(1)
            292 56-84-8P
            2542 110-17-8
L3   28 56-84-8P AND 110-17-8

⇐ S 9027-30-9 OR ASPARTASE ←————————————————————(2)
            168 9027-30-9
            147 ASPARTASE
L4   204 9027-30-9 OR ASPARTASE

⇐ S L3 AND L4 ←————————————————————————————(3)
L5   10 L3 AND L4

⇐ S 56-84-8P AND L4 ←————————————————————(4)
            292 56-84-8P
L6   42 56-84-8P AND L4

File 399:CA SEARCH 1967-1986 UD = 10406
(Corp. 1986 by the Amer. Chem. Soc.)

            Set Items Description
            - - - - - - - - - - - - - - - - -
?S RN = 56-84-8P AND RN = 110-17-8
            309 RN = 56-84-8P
            2853 RN = 110-17-8 (SEE ?IGNOTE)
    S1   29 RN = 56-84-8P AND RN = 110-17-8
?S RN = 9027-30-9 OR ASPARTASE
            167 RN = 9027-30-9
            153 ASPARTASE
    S2   210 RN = 9027-30-9 OR ASPARTASE
?S S1 AND S2
            29 S1
            210 S2
    S3   11 S1 AN S2
?S RN = 56-84-8P AND S2
            309 RN = 56-84-8P

210 S2

S4   45 RN = 56-84-8P AND S2 ←————————————(5)

(1) The Search command is used for combining the synthesis of L-aspartic acid (56-84-8P) and fumaric acid (110-17-8) using the Boolean operator AND. A sample of the results shows that many answers dealt with the chemical stereospecific synthesis of L-aspartic from fumaric acid.

(2) The Search command is used to search for the enzyme aspartase (9027-30-9) by combining the enzyme name and its CAS RN using the Boolean operator OR.

(3) Combining the two answer files (L3 and L4) using the Boolean operator AND results in ten relevant answers. Unfortunately it turned out that many known references on the subject were not included in these ten answers. Close examination of one of the 'non-retrieved' references showed that the enzymatic reaction was carried out on the sodium salt of fumaric acid.

(4) Assuming that aspartase acts on fumaric acid and its salts to produce L-aspartic acid, the synthesis of L-aspartic acid (56-84-8P) and aspartase (answer file L4) were combined using the Boolean operator AND to give 42 relevant answers.

(5) The same search strategy was used on Dialog File 399 to give 45 answers—42 relevant and 3 flash positive ones.

One should not be surprised that there is a difference in the number of results between the Dialog and the STN search; the difference in the 'synthesis algorithm' has been discussed earlier, but how could the difference in the citations of the RN of fumaric acid (2853 in Dialog, 2542 in STN) and aspartase (153 and 147, respectively) be explained?

The difference in the citation number for fumaric acid (110-17-8) between the Dialog System (2853) and the STN (2542) is caused by the fact that Dialog also retrieved undefined derivatives of fumaric acid (110-17-8D), while the STN retrieved only fumaric acid. Its undefined derivative retrieved separately and their combination gives identical results to the Dialog ones.

Dialog and STN have different name-segmentation algorithms. This difference is felt mainly in searching the Dictionary Files Basic Index (Files 300, 301, 328-331 on Dialog, the REG File on STN) [see C. Zigmund, *Database*, 8(3), 76 (1985)]. However, it also influences the bibliographic files; searching Dialog for aspartase also retrieved methylaspartase.

⇐ **FILE CA**
FILE 'CA' ENTERED AT 00:58:49 ON 03 MAR 86
COPYRIGHT 1986 BY THE AMERICAN CHEMICAL SOCIETY
⇐ **S 110-17-8 OR 110-17-8D**
            2542 110-17-8
            315 110-17-8D
L1   2853 110-17-8 OR 110-17-8D

⇐ **S METHYLASPARTASE OR ASPARTASE**
            8 METHYLASPARTASE
            147 ASPARTASE
L2   154 METHYLASPARTASE OR ASPARTASE

In cases where one cannot locate the desired synthetic information one should try and look for information on similar (analogous or homologous) reactions.

A bioorganic chemist is interested in the synthesis of DL-pentothenol [a compound better known by its *CA* Index Name—butanamide, 2-4-dihydroxy-N-(3-hydroxypropyl)-3,3-dimethyl-, (+-)] (16485-10-2) via the opening of DL-pantolactone [2(3*H*)-furanone, dihydro-3-hydroxy-4,4-dimethyl-, (+-)] (79-50-5) by 3-aminopropanol (156-87-6) (Reaction 9.3). No results were obtained in an online search of the *CA*. However, a search for the synthesis of the D-pantothenol (81-13-0) via the opening of the corresponding D-lactone (599-04-2) results in the retrieval of two procedures which in principle should work with the DL isomer without any problems.

$$HO(CH_2)_3NHCOCH(OH)\overset{CH_3}{\underset{CH_3}{C}}CH_2OH$$

(9.3)

← FILE CA
FILE 'CA' ENTERED AT 04:41:23 ON 25 JUL 86
COPYRIGHT 1986 BY THE AMERICAN CHEMICAL SOCIETY

← S 16485-10-2P AND 79-50-5 AND 156-87-6
        0 16485-10-2P
        56 79-50-5
        786 156-87-6
L1   0 16485-10-2P AND 79-50-5 AND 156-87-6

← S 81-13-0P AND 599-04-2 AND 156-87-6
        5 81-13-0P
        174 599-04-2
        786 156-87-6
L2   2 81-13-0P AND 599-04-2 AND 156-87-6

9.3. Designing a synthesis

When designing a synthesis one has to establish the best available reactions and procedures for combining together small or large fragments to form the desired skeleton, introducing functional group(s), etc., while taking into account the sensitivities of the existing parts of the molecule. Information about the various aspects of designing an organic synthesis can be found elsewhere (e.g. R.E. Ireland, *Organic Synthesis*, Prentice-Hall, Englewood Cliffs, N.J., 1969; S. Turner, *The Design of Organic Syntheses*, Elsevier, Amsterdam, 1976; N. Anand, J.S. Bindra, and S. Ranganathan, *Art in Organic Synthesis*, Holden Day, San Francisco, Calif., 1970; W. Carruthers, *Some Modern Methods of Organic Synthesis*, 2nd ed., Cambridge University Press, Cambridge, 1978; S. Warren, *Organic Synthesis: The Disconnection Approach*, Wiley, New York, 1983; R.O.C. Norman, *Principles of Organic Synthesis*, 2nd ed., Halsted Press, New York, NY, 1978; K.R. Mackie and D.M. Smith, *Guidebook to Organic Synthesis*, Longman, London, 1982; T. Lindberg, *Strategies and Tactics of Organic Synthesis*, Academic Press, New York, 1984).

Let us consider an organic chemist who is designing a multistep synthesis of a natural product, a key step in which is a reduction of a ketonic group to a secondary alcohol without affecting an α,β-unsaturated ketonic group (not the double bond nor the carbonyl). How could our organic chemist locate all available procedures for such a reduction (Reaction 9.4).

$$(9.4)$$

No useful information could be located using *Beilstein* or *CA*. *Beilstein* is a compendium of the properties of all organic compounds, and all the information is arranged according to the various chemical compounds. If our chemist had known a specific example of the reaction he is interested in he could have obtained the information concerning this specific reaction (the information would be found under the starting material and/or the product entry). The same is true searching the *CA*. Indeed, the *CA* General Subject Index includes classes of chemical compounds (e.g. ketones), chemical reactions (e.g. reduction), and so on. However, it includes studies of the compound as a class and studies of the reaction when the original document especially emphasizes the process or some aspects of it. Studies of specific compounds or reactions which are not emphasized as such in the primary source are located only under the headings of the specific compounds involved.

9.3.1. Organic class preparation

Our chemist should use one of the compendia of organic class preparations describing functional group preparations and/or transformations. Of the many avilable collections, a few that can be mentioned are: Volume 6 of *Methodicum Chemicum—Formation of C-N bonds*. *Compendium of Organic Synthetic Methods*, Wiley-Interscience, New York, 1971–84 (five volumes: the first two Volumes were compiled by I.T. Harrison and S. Harrison, the third Volume was compiled by L. Hegedus and L.G. Wade, and the last two Volumes were compiled by L.G. Wade). It is a systematic listing of functional group preparations arranged according to the type of transformation. The information is presented in the form of a reaction scheme followed by a reference without any accompanying text. Volumes 4 and 5 of the compendium are supplements covering new synthetic methods introduced in the years 1977–79 and 1980–82, respectively. J. Mathieu and J. Weill-Raynal, *Formation of C—C Bonds*, Georg Thieme, Stuttgart, 1973-1979 (three volumes). This is a presentation of the various reaction types which lead to formation of carbon-carbon bonds. Volume 1 deals with the introduction of the carbon atom bearing a functional group into a carbon skeleton (e.g. Reactions 9.5 and 9.6).

$$\geq C-H \longrightarrow \geq C-CH_2OH \qquad (9.5)$$

$$\geq C=C\leq \longrightarrow \geq C-\overset{|}{\underset{|}{C}}CH_2OH \qquad (9.6)$$

The second volume deals with attaching a carbon chain or an aromatic ring to a carbon skeleton (e.g. Reaction 9.7). The last volume deals with attaching a functional carbon chain to a carbon skeleton by specific types of reactions such as hydroxylation, acylation, tri- and tetramolecular reactions (reaction of a functional group with two or three molecules at the same time), insertion, and extrusion. S.R. Sandler and W. Karo, *Organic Functional Group Preparation*, Academic Press, New York, 1968–72 (three volumes). This is a collection of methods for the preparation of functional groups, each chapter bringing together the various reaction types for the introduction of the functional group. C.A. Buehler and D.E. Pearson, *Survey of Organic Synthesis*, Wiley, New York, 1970–77 (two volumes). This is a collection of methods for the preparation of functional groups. It states preferred methods, lists their limitations, and gives some reaction mechanisms. G. Hilgetag and A. Martini (Editors), *Weygand/Hilgetag Preparative Organic Chemistry*, Wiley, New York, 1972. This is an English translation of the 4th German Edition of the compendium. The material is arranged on the basis of the formation of various carbon bonds. There is a detailed table of contents which enables one to locate the desired chemical transformation one is seeking. The material is arranged in four parts—reaction on the carbon skeleton, formation of carbon-carbon bonds, cleavage of carbon-carbon bonds, and rearrangements of carbon compounds. The book includes three appendices dealing with purification and drying of organic solvents, preparation and purification of gases, and preparative organic work with small amounts. Most of the books have experimental procedures for the various reaction types. However, in all cases the emphasis is on the specific chemical transformation involved, with little (if any) consideration of other functional groups which are not involved in the reaction itself.

$$\gtreqless C-H \quad + \quad Ar \quad \longrightarrow \quad \gtreqless C-Ar \qquad (9.7)$$

However, our chemist should try and use W. Theilheimer, *Synthetic Methods of Organic Chemistry*.

9.3.1.1. *Synthetic Methods of Organic Chemistry*

Synthetic Methods of Organic Chemistry is an annual publication devoted to new synthetic methods of organic compounds, improvements of known methods, and the collection of old proven methods scattered throughout the literature. Its first four volumes were published in German (1946–49). Volumes 1 and 2 have been translated into English, while an English Keywords Index was issued for Volumes 3 and 4. From Volume 5 onwards it is published in English.

Each volume describes the newly developed or improved synthetic methods reported in the preceding year, the material being arranged according to the reactions involved. These reactions are indexed by 'Reaction Symbols' developed ty Theilheimer. The first part of the Reaction Symbol refers to the chemical bond formed during the reaction. The order in which the elements are listed is H, O, N, Hal, S, remaining elements (Re), C. The second part is the method by which the bond is formed (e.g. rearrangement, exchange). The last part refers to the bond destroyed in the reaction or to the element that is eliminated. Thus the reduction of the carbonyl group to a secondary alcoholic group would be described as HC⇂OC. A table of contents of the Reaction Symbols reported in a given volume is given under the title 'Systematic Survey'. A method is subdivided on the basis of reagents used. At the end of each volume one can find the volume indexes. The Subject Index gives names of methods, types of compounds and reagents. However, it does not give specific compound names or author names. These can be found in the *CA* Indexes or other abstracts journal Indexes. The Formula Index does not give a page or a reaction number, but a formula for functional combination together with the names of compounds whose synthesis has been described in the volume (elements appear in alphabetical order except for carbon, which appears last; Hal is used for all halogens and hydrogens and Rs are omitted). There is also a section entitled 'Supplementary References', which serve to update the literature on methods reported in earlier volumes.

Let us go back to our organic chemist and see what he could find using this source. First he checked the Systematic Survey in order to find the location of the desired reaction type. The Systematic Survey of the latest available volume (Volume 39, 1985) gives information for Volumes 36–39. No useful reaction could be located in those five Volumes under the Reaction Symbol HC⇓OC; the same was true for Volumes 33–35. In Volume 32 (1978) our chemist found on page 19 (reaction 27) an answer he was looking for. This is a preferential *in situ* Oppenauer oxidation of the corresponding aluminium alkoxide obtained by reacting 4-androstene-3,17-dione (**9.9**) with 20% solution of diisobutylaluminium hydride in toluene to give testosterone (**9.10**) in 79% yield [U. Eder, *Ber.*, **109**, 2954 (1976)]. (Reaction 9.8).

9.9   9.10   (9.8)

Our chemist decided to do a citation search of Eder's article and found that the above paper was cited five times during the years 1976-85. Of those five articles one was a review article, another describes the selective reduction of aldehydes to primary alcohols in the presence of another aldehydic group, and three dealt with the desired reaction. Two of the articles describe usage of the preferential *in situ* Oppenauer oxidation—the reduction of 4-androstene-3,17-dione-19(2′-tetrahydropyranyl ether) (**9.11**) to the corresponding testosterone ether (**9.12**) [P.N. Rao *et al*. *J. Ster. Biochem.*, **9**, 539 (1978)] and the reduction of *B*-norandrost-4-ene-3,17-dione (**9.13**) to *B*-norandrost-4-ene-17-beta-ol-3-one (**9.14**) [H.L. Holland, *Canad. J. Chem.*, **59**, 1651 (1981)]. The last paper—E. Dincan *et al. Tetrahedron*, **38**, 1755

9.11   9.12

184

(1982)—describes a new method for the selective reduction of a ketonic carbonyl in the presence of an unsaturated ketonic group. The reduction is carried out using tetrabutylammonium borohydride in methanol; androst-4-ene-3,17-dione (9.9) is reduced in over 90% to testosterone (9.10) and compound 9.15 is obtained in the same yield by the reduction of progresterone (9.16).

9.13

9.14

9.15

9.16

9.3.1.2. Chemical Reactions Documentation Service

Our chemist was wondering if such a search could be carried out online, thus enabling him to get a much better coverage of the last 18–24 months as well as shorten and simplify the manual search. Indeed the search could be carried out online using the Derwent data base *Chemical Reactions Documentation Service* (*CRDS*) which is available on Orbit. It covers the information recorded in *Synthetic Methods of Organic Chemistry* as well as the data reported in the *Journal of Synthetic Methods* since their inception (1942 and 1975, respectively). The online search of our chemist's problem results in retrieval of seven references.

Examination of the search results showed that only four out of the seven references were relevant ones. There was an entry to Eder's paper [*Ber.*, **109**, 2954 (1976)] which our chemist retrieved manually earlier. The other entries were B.J. Hussey *et al.*, *Tetrahedron*, **38**, 3769 (1982); F.M. Menger *et al.*, *J. Org. Chem.*, **45**, 2724 (1980); G. Paolucci *et al.*, *J. Chem. Soc. Perkin II*, 1129 (1979).

Hussey uses zinc bistetrahydroborate/DMF complex $(Zn(BH_4)_2).1.5DMF$ as the reducing agent. The paper reports the preferential reduction of 9.9 to 9.10, of 9.16 to 9.15, and of 9-methyl-delta-5(10)octalin-1,6-dione (9.17) to 9-methyl-delta 5(10)octalin-1-ol-6-one (9.18). All reductions proceed with over 80% yield. Menger shows that a polymeric reagent obtained by reacting poly 2-vinyl pyridine with borane and methyl sulphide complex in THF does not react with cyclohex-2-en-1-one (**9.19**) but reduces cyclohexanone (**9.20**) to cyclohexanol (**9.21**) in excellent yield. Paolucci shows that nickel dihydrobis(pyrazol-1-yl)-borate reacts in a similar way to the poly 2-vinyl pyridine borane compound (Menger's polymer), thus reducing **9.20** to **9.21** while not reacting at all with compound **9.19**.

9.17 **9.18** **9.19**

9.20 **9.21**

Citation searching of Hussey, Menger, and Paolucci works did not retrieve any more relevant answers. At that stage our chemist had four different reagents to choose from.

The *CRDS* search is based on using various reactions and chemical codes and combining them by the various Boolean operators. However, there are various reaction search systems based on graphic input, e.g. REACCS (Reaction Access System) developed by MDL, RMS (Reaction Management System) developed by Telesystems Darc, SYNLIB (Synthetic Library) developed by Distributed Chemical Graphics.

9.3.1.3. Synthetic Library (SYNLIB)

Our chemist decided to search for the selective ketonic reduction (Reaction 9.4) using SYNLIB. SYNLIB is a software package for managing and searching chemical reactions data bases inhouse. The program at present includes a reaction library of about 32,000 organic reactions (the reaction number grows all the time). Searches are based upon chemical structure, reagents, reaction conditions, author, and/or references.

The first step in the SYNLIB search is drawing the two structures in the input mode (Fig. 9.1). The next step is to enter the constraint mode and at RX CNDX select '*REQU', '*W', 'REDU'. RREDU indicates that the search is limited to those entries entered as reduction. At BOND select 'BREAK' and 'C=0'. The appearance of BC=0 means that the program will select those transformations that break (remove) a C=0 from the starting material (Fig. 9.2). The next step is to return to the input mode (via RTRN) and select SRCH for searching. On the Select Search screen (Fig. 9.3) one selects 'BROAD SEARCH'.

186

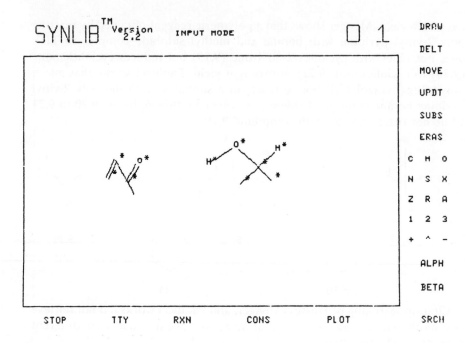

Figure 9.1 SYNLIB draw mode

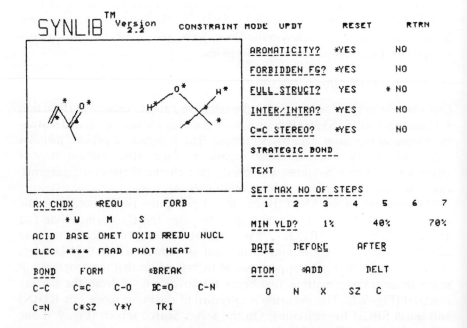

Figure 9.2 SYNLIB constraints mode

187

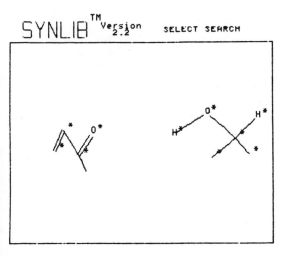

SYNLIB™ Version 2.2   SELECT SEARCH

CONSTRAINTS ACTIVE

BROAD SEARCH

NARROW SEARCH

DIRECTORY ADDRESS

JOURNAL REFERENCE

RETURN

Figure 9.3   SYNLIB search mode

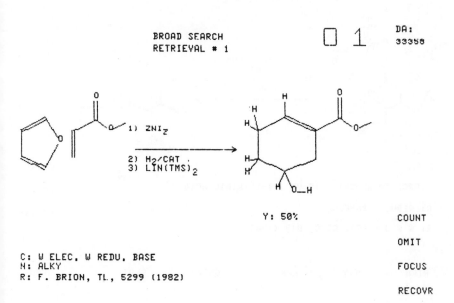

BROAD SEARCH
RETRIEVAL # 1

□ 1   DA:
33358

1) ZNI₂

2) H₂/CAT
3) LIN(TMS)₂

Y: 50%

C: W ELEC, W REDU, BASE
N: ALKY
R: F. BRION, TL., 5299 (1982)

COUNT

OMIT

FOCUS

RECOVR

Figure 9.4   SYNLIB answer 1

188

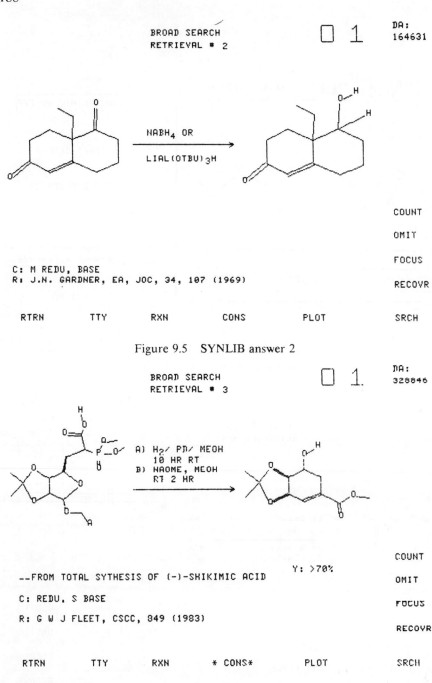

NABH₄ OR

LIAL(OTBU)₃H

C: M REDU, BASE
R: J.N. GARDNER, EA, JOC, 34, 107 (1969)

Figure 9.5   SYNLIB answer 2

A) H₂/ PD/ MEOH
   10 HR RT
B) NAOME, MEOH
   RT 2 HR

Y: >70%

--FROM TOTAL SYTHESIS OF (-)-SHIKIMIC ACID

C: REDU. S BASE

R: G W J FLEET, CSCC, 849 (1983)

Figure 9.6   SYNLIB answer 3

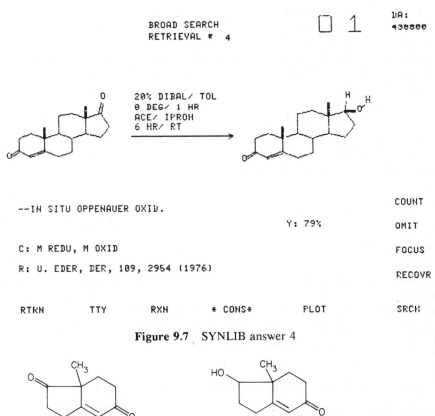

Figure 9.7 SYNLIB answer 4

There are only four hits: the first hit has a directory address 33358 (Fig. 9.4), the second 164631 (Fig. 9.5), the third 328846 (Fig. 9.6), and the fourth 438800 (Fig. 9.7). Examination of those answers shows that only two answers are relevant. One (438800) is Eder's paper [*Ber.*, **109**, 2954 (1976)], which has been the only manual retrieval using Theilheimer's *Synthetic Methods*. The other one (164631) is a paper by Gardner *et al.* [*J. Org. Chem.*, **34**, 107 (1969)] which uses sodium borohydride or lithium aluminium *t*-butoxide for the preferential reduction of the ketonic group (see Fig. 9.5). A citation search of Gardner's work results in locating two other examples using sodium borohydride for the preferential reduction of **9.22** to **9.23** [*J. Org. Chem.*, **40**, 675 (1975)], and of **9.24** to **9.25** [*J. Org. Chem.*, **42**, 2754 (1977)].

190

9.3.1.4. Organic Reactions Accessed by Computer (ORAC)

Organic Reactions Accessed by Computer (*ORAC*) is another computer program for reaction indexing and retrieval; at the end of 1987 it contained about 40,000 reactions. Searching ORAC for the preferential reduction of a carboxyl group to an hydroxyl group is presence of α,β unsaturated carboxyl (reaction 9.4) results in 6 hits.

Two of the hits were retrieved before: one is Eder's paper (*Ber.*, 109, 2954 (1976)) which describes the preferential *in situ* Oppenauer oxidation of 9.9 to 9.10; the other is Hussey *et al.* paper (Tetrahedron, 38, 3769 (1982)) which describes the use of zinc *bis*-tetrahydroborate/DMF complex as a reducing agent.

Two other hits are D.H.R. Barton *et al.* work [*Chem. Comm.*, 1072 (1972); *J. Chem. Soc. Perkin Trans. I.*, 1075 (1975)] in which they describe the selective formation of an enolate anion as a protection for a carbonyl group against a reduction of LiAlH$_4$ (reaction 9.9).

The other two hits are the reduction of prednisone BMD (**9.26**) to prednisolane BDM (**9.27**) by lithium tri-*t*-butoxyaluminium hydride. (J.A. Zderic and J. Iriarte, *J. Org. Chem.*, **27**, 1756 (1962)) and the reduction of compound **9.28** to the 5-β alcohol (**9.29a**) by NaBH$_4$.CeCl$_3$ or to the 5-α alcohol (**9.29b**) by NaBH$_4$ in a mixture of CH$_2$Cl$_2$/iPrOH (W.C. Still and V.J. Novack, *J. Am. Chem. Soc.*, **106**, 1148 (1984)).

9.3.1.5. Methoden der Organischen Chemie (Houben-Weyl)

Let us consider a peptide chemist who is designing a synthesis of some alpha- and gamma-glutamyl peptides. He would like to acquire a general overview concerning the preferential protection of the alpha- and gamma-carboxylic groups of glutamic acid. He discussed this with some colleagues who suggested that he should read a book dealing with peptide syntheses. He read the recommended book and felt that he would like to obtain some more detailed information concerning the specific reaction conditions of the introduction and/or removal of the carboxylic protecting groups. Another colleague who heard about his problem referred him to *Methoden der Organischen Chemie* (Houben-Weyl).

Methoden der Organischen Chemie was compiled by Th. Weyl at the beginning of the century and was first published in 1909, entitled *Die Methoden der Organischen Chemie: ein Handbuch fur die Arbeiten in laboratorium*. The last edition is the fourth edition which was published in 16 volumes and 66 parts during the years 1952–86. In 1982 the publication of Supplements and Additions was started to the fourth edition (*Eeweiterungs and Flogeband zur Vierten Auflage*). It is an open-end publication; its plans at present call for the publication of 22 volumes in an undisclosed number of parts. The publication was intended throughout all its editions to deal comprehensively and critically with all types of experimental methods and to present the

$$\text{Ph}_3\text{CLi;LiAlH}_4/\text{THF} \qquad (9.9)$$

9.26

9.27

9.28

9.29

a = 5 β–hydroxy
b = 5 α–hydroxy

theoretical background of each method, followed by pertinent examples with full experimental details. In order to produce the various volumes as early as possible, volumes were and are published as soon as they are completed, regardless of their sequential order. Thus different parts of the same volume may appear a few years apart, so that there is a difference in their coverage period. One can locate the desired information by using the table of contents, which are very detailed, or by using the subvolumes (parts) Subject Index. The chemical compound names in the Subject Index are given in the traditional German manner and not according to IUPAC nomenclature. If one is trying to locate a chemical compound using the Subject Index, one should reconstruct the German name (e.g. calculating the molecular formula of the desired compound and consulting the *Beilstein* Formula Index for the 'right' name).

Our peptide chemist became aware that a whole volume (Volume 15) of *Methoden der Organischen Chemie* is devoted to peptide synthesis. On examining the two parts of Volume 15, it was clear that the required information is available in the first part of the Volume. Indeed, on going through the table of contents one can locate Section 36.600, Carboxy Funktion, pp. 634–701. These 68 pages cover all available information up to the beginning of 1974, followed by many experimental procedures. Any further search from 1974 onwards would have to be carried out using the *CA* or the *SCI*. Another approach would be to try to locate a very recent review article dealing with that particular subject.

Methoden der Organischen Chemie could also be used to locate some specific synthetic information. For example, a chemist dealing with organic nitrogen compounds may be looking for a simple procedure for the nitration of aliphatic hydrocarbons. He remembered that when he took the laboratory course in organic chemistry his instructor said something about using metal nitrates as nitrating agents for aliphatic hydrocarbons. He decided to try to locate the information using *Methoden der Organischen Chemie*. He turned to Volume 10/1—Nitro, Nitroso, and Hydroxylamino Compounds. On going through the table of contents he found Section A1a/24—Nitration with Metallic Nitrates, P. 37. There he found some discussion about the usage of various metal nitrates as nitrating agents as well as two experimental procedures, the nitration of cyclopentane to nitrocyclopentane by $Al(NO_3)_3$ and the nitration of ethyl acetate to ethyl nitroacetate by $Cu(NO_3)_2$ followed by a few references. (It is interesting that out of the various compendia of organic class preparations mentioned earlier, this nitration reaction is reported only in *Methodicum Chimicum*. In Volume 6, *Formation of C—N Bonds* there is a remark about the use of metallic nitrates for the nitration of aliphatic hydrocarbons, followed by a reference to *Methoden der Organischen Chemie*).

9.3.1.6. Perspectives

The present situation as well as perspectives in chemical reaction searching were discussed in a special symposium organized by the Chemical Structure Association at the University of York on 8–11 July 1985. The proceedings of the symposia were edited by P. Villett and were entitled *Modern Approaches to Chemical Reaction Searching* (Gower, Hants, England, 1986).

9.3.2. Information on reagents and solvents

During the design of a synthesis a chemist will need information about the reagents and solvents he is going to use, e.g. their preparation (if not commercially available), properties, and purification. An encylopaedic source of this kind is L.F. Fieser and M. Fieser, *Reagents for Organic Synthesis*. The reagents described are mainly organic but some inorganic compounds are also included. It is a multi-volume publication (12 volumes have been published up to date), in which new volumes include, in addition to new reagents, supplementary material about reagents reported earlier.

There are various serial publications devoted in whole or in part to synthetic reagents. Wiley has been publishing for years a serial entitled *Synthetic Reagents* edited by J.S. Pizey. In the six volumes published one may find monographs on reagents such as: chloramine T, H_2O_2, H_3PO_3, NH_3, ICl, $Ta(CH_3COO)_3$ and $Ta(CF_3COO)_3$, $Hg(CH_3COO)_2$, HIO_4, DDQ, I_2, $Pb(CH_3COO)_4$, NBS, CH_2N_2, MnO_2, Raney nickel, DMF, $LiAlH_4$, HgO, $SOCl_2$. Springer Verlag is publishing a serial entitled *Reactivity and Structure*; some of the volumes in this serial publication deal with groups of reagents in organic synthesis [e.g. Volume 4, *Phase Transfer Catalysis in Organic Synthesis* (W.P. Weber and G.W. Gokel); Volume 10, *Organic Synthesis with Paladium Compounds* (J. Tsuji); Volume 14, *Silicon Reagents for Organic Synthesis* (W.P. Weber); Volume 22, *Organic Mercury Compounds in Organic Synthesis* (R.C. Larock)]. Academic Press recently started the publication of *Best Synthetic Methods* of which two volumes have been published: R.F. Heck, *Paladium Reagents in Organic Synthesis* and Haines' *Methods for the Oxidation of Organic Compounds*. Information about the usage of organometallic compounds as synthetic reagents and catalysts can be found in Volumes 7 and 8 of *Comprehensive Organometalic Chemistry* which was edited by G. Wilkinson. It is a nine-volume work covering the various aspects of organometallic chemistry: synthesis, reactions, and structure of organometallic compounds. (It resembles in its coverage the companion works— *Comprehensive Inorganic Chemistry* and *Comprehensive Organic Chemistry*.) The first six volumes give a detailed treatment of the organic chemistry of the main groups and the transition elements. The last volume (Volume 9) is an index volume which besides the usual indexes (Subject, Formula, Author) contains also Index of Structures by Diffraction Methods of all organometallic structures studied and Index of Review Articles and Books covering the entire organometallic literature up to 1981.

Where could a synthetic chemist locate information about the purification and drying of organic solvents? How would one purify and dry acetone? Some information could be found in Fieser and Fieser's *Reagents for Organic Synthesis*. Acetone is not reported in Volume 1 but it is discussed in Volume 2. One can find there some general statements about purifying acetone by refluxing it with small amounts of permanganate until the purple colour persists, distillation, drying over anhydrous K_2CO_3 for two days and filtering, followed by a second distillation. This is all of the available information, which in certain cases should be good enough but in other cases could be completely unsatisfactory. Similar information could be found in *Weygand/Hilgetag Preparative Organic Chemistry*. The best available information is to be found in Volume II of A. Weissberger's *Techniques of Chemistry—Organic Solvents*. This volume includes in addition to a collection of various detailed procedures for the purification of many organic solvents, an extensive collection of their physical properties, thus enabling the scientist to choose the appropriate solvent for a specific application.

An important source for information about the purification of solvents and reagents is D.D. Perin, W.L.F. Amarego, and D.R. Perin's second edition of *Purification of Laboratory Chemicals* (Pergamon Press, 1983). It brings information about the purification and/or repurification of about 4000 individual chemicals and biochemicals (about 3200 organic compounds and about 800 inorganic and metalloorganic compounds) as well as general methods for the purification of 28 classes of compounds.

9.3.2.1. Techniques of Chemistry

Techniques of Chemistry, which was published since 1970, is a collection of 18 volumes in 35 parts dealing with various methods, techniques, and operations used in chemistry. There is a detailed description of the various methods and operations, including a theoretical background. Many of the techniques described in the series are described in many other sources (e.g. NMR Spectroscopy, IR Spectroscopy) but others are not so easily located (e.g. Organic Solvents, Stereospecific Synthesis).

This series supercedes two other series edited by A. Weissberger and published during the 1950s and 1960s— *Techniques in Organic Chemistry* and *Techniques in Inorganic Chemistry*. Indeed many of the same techniques are used today in both organic and inorganic chemistry.

9.4. Getting an overview

A synthetic chemist (especially a young one) may need from time to time, like any other scientist, to acquire a general overview of a specific problem or area. Further, he may have to find the optimum conditions for the reactions he is using or planning to use, and an accurate conception of those reactions. All this needs an extensive literature search. In this respect a synthetic organic chemist has a clear advantage over colleagues who deal with inorganic synthesis. There are compilations that give an overview of various organic reactions. One of those compilations has been discussed before— *Methoden der Organischen Chemie* (Houben-Weyl). Others are Patai's series *The Chemistry of the Functional Groups, Organic Reactions*, etc. Unfortunately there are no similar publications devoted to inorganic chemistry.

We have already seen an example of getting a general overview—in the case of the peptide chemist dealing with the synthesis of alpha- and gammaglutamyl peptides (p. 190). He was using a book devoted to peptide synthesis (e.g. M. Bodanszky and A. Bodanszky, *The Practice of Peptide Synthesis*, Springer Verlag, 1984). Read the appropriate section in Volume 15/1 of *Methoden der Organischen Chemie* as well as conduct an extensive literature search of the last few years.

9.4.1. *The Chemistry of the Functional Groups*

The Chemistry of the Functional Groups edited by S. Patai and published by Wiley is a series of about 30 volumes, each volume covering all aspects of the chemistry of one chemical functional group. The emphasis is on the chemical and physical properties of the functional group itself, and the influence of the group upon its immediate vicinity as well as on the behaviour of the whole molecule. Supplementary volumes have been published and are planned for the various volumes.

Although the series deals with all aspects of organic chemistry it is of special interest to the synthetic organic chemist; e.g. 60% of the material in the carbon-carbon triple bond volume is of importance to the synthetic organic chemist. One can find there coverage of the following synthetic subjects: application of acetylenes in organic synthesis, electrophilic addition to the $C \equiv C$ bond, nucleophilic attacks on acetylenes, photochemistry of the $C \equiv C$ bond, and cycloadditions and cyclizations involving triple bonds.

9.4.2. *Organic Reactions*

Organic Reactions is a collection of monographs devoted to single reactions in organic chemistry. Sometimes it even deals with only a definite phase of a reaction of a wide applicability. The subject is presented from the preparative viewpoint, with emphasis on the potential and limitations of the reactions discussed, interfering effects, etc. Each monograph (Chapter) is accompanied by tables including all the examples of the reactions known to the author. The Editorial Board was aware of the problem of updating the various reports. Since Volume 22, updating of selected subjects has been presented as brief reports (not necessarily complete) covering the literature since the earlier report; e.g. Clemensen reduction (zinc/hydrochloric acid) is covered in Volume 1 and updated in Volume 22. Each of the volumes contains a detailed Subject Index to itself and a cumulative Author and Chapter Index to the whole series.

CHAPTER 10

Structural and Substructural Searches

Many of the search examples discussed in Chapters 6, 7, and 9 are structural search problems (e.g., melting point of compound **6.1**, mass spectra of compound **6.43**, synthesis of compound **9.3**, production of benzoic acid by carboxylation of benzene). In other cases one is in need of information related only to part of the molecule; a search for such information is known as a substructural search. An example of a substructural search is: what are the biological properties of compounds having five-membered rings with structure **10.1** or **10.2**. Which search tools are available for such searches and how could they be used?

10.1	**10.2**

Efficiency in such searches depends upon language and terms used. The structural formula has become the common, pervasive, and useful language in structural chemistry. Indeed, the main problem in conducting a structural or substructural search is the conversion of a three-dimensional structure (or a two-dimensional structure with known stereochemical characteristics) into a one-dimensional representation (e.g. a molecular formula, a chemical name, a notation, a connection table). Once such a conversion has taken place, one should use the various manual and/or computerized indexes as tools in one's search.

10.1. Structural search

A structural search is usually carried out with the aid of a formula index or a chemical name (or chemical substance) index. Each of the various information sources has its own index system. Although the various index systems are similar, some differences exist which the user should be aware of.

196

10.1.1. Formula indexes

A molecular formula index, empirical formula index, or just formula index, links the molecular formula of a chemical substance to the page number (or the abstract number) in which it was discussed (or in which it was reported). One has to keep in mind that usually more than one substance is reported under a molecular formula entry (in the 10th Collective Formula Index of the *CA* there are eight compounds having the molecular formula C_3H_8O, e.g. methoxyethane, $CH_3OCH_2CH_3$; 1-propanol, $CH_3CH_2CH_2OH$; 2-propanol, $CH_3CHOHCH_3$. Thirty-six compounds having the molecular formula C_4H_8O, e.g. butanal, $CH_3CH_2CH_2CHO$; 2-butanone, $CH_3COCH_2CH_3$; *cis-* and *trans*-2-buten-1-ol, $CH_3CH=CHCH_2OH$; 1-methoxy-1-propene, $CH_3CH=CHOCH_3$. The formula entry is usually followed by the chemical substance name, the various substances having the same formula being arranged alphabetically under the formula entry. In many cases the compound index name is known to the searcher and he can use it while searching the formula index [he may use it while working with the chemical name (or substance or compound, etc.) index]; in other cases the searcher can locate and identify the compound name while working with the formula index. However, there are cases in which one cannot use the formula index without knowing the compound name. For example, an undergraduate student was interested in knowing if anything had been reported about compound **10.3** during the years 1972–76. He first calculated the molecular formula of **10.3**, $C_{12}H_8O_3$, then turned to the 9th Collective Formula Index of *CA*. There under $C_{12}H_8O_3$ he found entries to 49 different compounds. Although none of those compounds seemed to be the one he was looking for, he wanted to be sure about this conclusion and decided to try and find the chemical name of the compound (see p. 201).

10.3

Most modern formulae indexes are arranged according to the Hill System (see p. 91). However, two other systems were very widely used during the 19th and the beginning of the 20th Century—the Richter System (see p. 91) which covers organic compounds and the Hoffman System which covers inorganic compounds. The Hoffman System is arranged according to the priority numbers of the elements in such a way that the highest priority numbers are always placed first. The priority numbers were selected by Hoffman himself and are similar, but not identical with the System Numbers in *Gmelin*. In that system sulphuric acid, for example, is indexed as SO_4H_2. There are also cases in which the formula index is arranged in a completely different form, e.g. *Gmelin's Handbuch der Anorganischen Chemie* (see p. 103). Before one uses a formula index, one should be sure that one knows the system on which the index is built.

There are formula indexes in which the various chemical names which appeared under the empirical formula are followed by descriptive phrases. These phrases give information about the context in which the compound is reported, thus enabling the searcher to determine immediately the relevance of the entry to his specific problem (e.g. the Formula Index in *Gmelin*). On the other hand, there are formula indexes which do not give any descriptive phrases; in such a case one has to check each of the various entries in order to determine their relevancy (e.g. *CA* Formula Index).

As a rule, when one is dealing with a structural search one should preferably use formula index which includes descriptive phrases rather than a chemical name index; a formula index which does not have any descriptive phrases is still preferable to a chemical name index without such phrases. On the other hand, one should preferably use a chemical name index which includes descriptive phrases rather than a formula index without any such phrases. Indeed, we have seen earlier (p. 148) the advantage of using the *CA* Chemical Substance Index and/or *CA* Subject Index over the corresponding Formula Index.

One should be aware of the existence of the various modified formula indexes (e.g. Rotated Formula Index) and use them whenever such a usage has an advantage over using a regular formula index.

10.1.2. Chemical name indexes

A chemical name index or chemical substance index links the names of compounds to the pages on which they have been discussed or to the corresponding abstracts numbers in which they were referred to. There are indexes which include chemical compounds only (e.g. *CA* Chemical Substance Index, which has been published since 1972, and *Organic Syntheses* General Index), whereas others include all subjects reported—not necessarily chemical compounds (e.g. *CA* Subject Index prior to 1972 and Houben-Weyl's Volume Indexes).

The main problem in the arrangement of a chemical substance index is the naming of the various compounds. A chemical compound usually has more than one name; e.g. even a simple compound such as $NH_2CH_2CH_2OH$ has at least six different names: 2-aminoethanol; 2-aminoethylalcohol; 1-amino-2-hydroxyethane; ethane, 1-amino-2-hydroxy-; 2-hydroxyethylamine; beta-hydroxyethylamine. One should keep in mind that among the over 8,000,000 compounds stored in the CAS data base there are 17 compounds having more than 150 names each, the record is held by polyethylene, which is identified by 945 different names. There are also cases in which more than one compound is called by the same name (e.g. compounds **8.1** and **8.2** are named dioxin; see p. 160). TCP is a well known compound—2,4,5-trichlorophenol (**8.4**); however there are four other compounds having the same synonym—2,3,4,6-tetrachlorophenol (**10.4**); piperidine, 1-(1-(2 thienyl)cyclohexyl)—(**10.5**); 2,2,3-trichloroproprionic acid (**10.6**); phosphoric acid, tris(methylphenyl)ester (**10.7**).

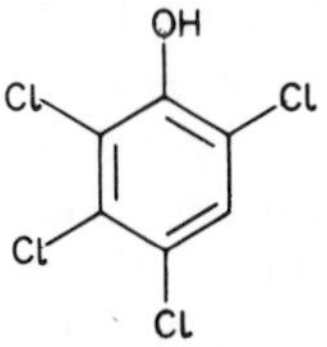

10.4

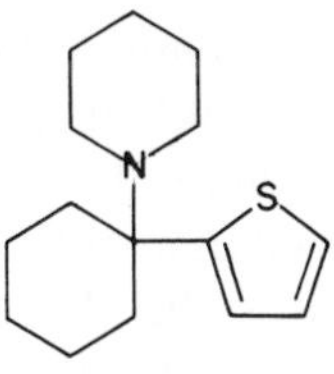

10.5

CLCH$_2$CCL$_2$CO$_2$H

10.6

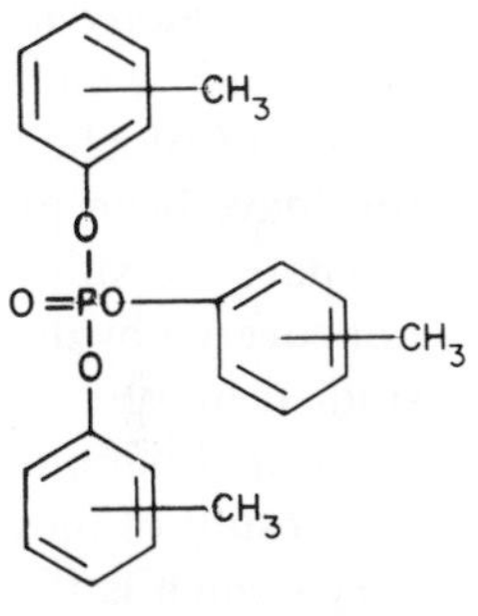

10.7

When one is retrieving information about a chemical compound using a chemical name (or chemical substance) index one cannot (and definitely should not) search under all the names that have been used for the compound in the various publications. One should be able to locate all the information concerning a specific compound under one entry even when different names were used for that compound in various sources (e.g. information about chlorpromazine, 2-chloropromazine, aminazine, and propaphenin should be found under one heading entry, and indeed they are all indexed in the *CA* under 10*H*-phenothiazine-10-propanamine,2-chloro-*N*,*N*-dimethyl-, which is the *CA* Index Name of the compound).

One would expect that a name used in a chemical name index should reveal the whole structure of the molecule. Furthermore, the index itself should try to bring together as much as possible closely related compounds. How could these two demands as well as the assurance of using one accepted name for each chemical compound in the chemical name index be achieved? It could be achieved by naming the various compounds according to accepted nomenclature rules, rules which are very detailed and cover every atomic combination.

Work has been and is still being carried out in order to build and develop a unified nomenclature system. This task has been undertaken by IUPAC throughout its various Nomenclature Committees and Commissions. Basic texts which describe and explain some of the basic rules for the most important types of chemical compounds are: R.S. Chan and O.C. Dermer's *Intro-*

200

duction to Chemical Nomenclature, 5th ed., Butterworth, London, 1979; R. Lees and A. Smith's *Chemical Nomenclature Usage*, Ellis Horwood, Chichester, 1983. IUPAC recommendations and detailed rules for naming chemical compounds have been compiled in three books: *Nomenclature of Organic Chemistry*, Pergamon Press, Oxford, 1979, which is also known as the 'Blue Book'; *Nomenclature of Inorganic Chemistry*, Pergamon Press, Oxford, 1971, which is also known as the 'Red Book' (it has a companion guide *How to Name an Inorganic Substance*, Pergamon Press, Oxford, 1976); *Compendium of Macromolecular Nomenclature*, Blackwell Scientific Publications, London, 1985, which is also known as the 'Purple Book'. The International Union of Biochemistry (IUB) recommendations for biochemical nomenclature have been published in a book entitled *Biochemical Nomenclature*.

There are still many types and classes of chemical compounds which are not covered by definite IUPAC rules. Their systematic nomenclature is being studied by specialized commissions, and the results of these studies will be additional definite rules, tentative rules, or just naming recommendations. The results of these studies are reported regularly in *Pure and Applied Chemistry* (which is the official organ of IUPAC) and in the *IUPAC Information Bulletin*.

A comprehensive compilation of all chemical nomenclature rules and/or recommendation publications which have been published in recent years is published from time to time in *Chemistry International*. The last two lists have been published in October 1981 [*Chem. Int.*, (5)28, (1981)] and in October 1982 [*Chem. Int.* (5)15, (1982)].

The chemical nomenclature used by the *CA* is very similar, but not identical, with the IUPAC system. It has been developed over the years in parallel and more or less in accordance with the development of the IUPAC rules. One has to keep in mind that the objectives of *CA* and IUPAC, although similar, are not identical, as chemical names best suited for indexing are not always best suited for other purposes. A major revision of the *CA* Index Names was carried out at the beginning of the 9th Collective Index period (1972). These *CA* Index Names have been used unchanged since then (9th–12th Collective Index periods). Trivial and non-systematic names which were often used prior to 1972 are no longer used and only a very few non-systematic names are now in use. Some of the changes that have taken place are as follows:

Prior to 1972	*Since 1972*
Aniline	Benzeneamine
Anisole,*p*-methyl-	Benzene,1-methoxy-4-methyl-
Bromobenzene	Benzene,bromo-
Nitrobenzene	Benzene,nitro-
Ethylene	Ethene
Ethylenimine	Aziridine
Ethylene oxide	Oxirane
Ethylenediamine	1,2-Ethanediamine

The *CA* naming policy is explained in the *Chemical Substance Name Selection, Manual for the 9th Collective Period*, which was published in two volumes by CAS in 1973.

The basic *CA* nomenclature rules for organic compounds can be found in J.E. Banks' *Naming Organic Compounds, A Programmed Introduction to Organic Chemistry*, 2nd ed., W.B. Saunders, Philadelphia, 1976. More detailed information about selecting *CA* Index Names can be found in Appendix IV to the latest edition of the *Index Guide, Selection of Index Names for Chemical Substances*. A reprint of the appendix is available for a nominal fee from the Marketing Department of CAS.

Chemical name indexes usually use either the IUPAC nomenclature or the *CA* Index Names. One should know in advance, before using a chemical index name, the nomenclature system on which it is based.

Let us go back to our undergraduate student—how could he determine the *CA* Index Name of compound **10.3**? One route is to use one or two of the sources which have been discussed in this section in order to determine the *CA* Index Name. In this particular case it may turn out to be a long and difficult task. However, as **10.3** is a cyclic compound our student should be able to obtain results easily and quickly by using the *Ring System Handbook*. If he could name the ring system **10.8** he could easily name the compound itself.

10.8

10.1.2.1. Ring System Handbook

The *Ring System Handbook* (which superseded the *Parent Compound Handbook*) is a reference tool for the identification of ring systems. The 1984 edition is composed of two parts, the Ring System File (RSF), containing about 60,000 ring systems and 179 cage systems (in two volumes), and the indexes—the Ring Formula Index (RFI) and the Ring Name Index (RNI). A cumulative supplement to the *Handbook* is published semiannually. The Ring System File includes for each ring system its Ring File (RF) number, which serves as an access to the specific ring system only in the specific edition or supplement it is reported in, CAS RN, structural diagram of the ring system illustrating its numbering system, the current *CA* Index Name of the ring, and its Wiswesser Line Notations. The ring systems are arranged in the RF according to their ring analysis.

The first step in using the *Ring System Handbook* is to find the RF number of the parent compound **10.8**, which is done by using the Ring Formula Index. The RFI is a molecular formula index arranged according to the Hill

System excluding hydrogens, further divided according to the number of rings and the ring formulae (without hydrogens). The parent compound has the molecular formula $C_{10}H_6O$ ($C_{10}O$ excluding hydrogens) with four rings (a four-ring system), having a ring elemental analysis of C_4O,C_5,C_5,C_5. Our student found in the RFI six compounds having a formula of $C_{10}O$, being a four-ring system and having a ring elemental analysis of C_4O,C_5,C_5,C_5. These compounds have the following RF numbers: 30671, 30679, 30681, 30684, 30685, and 30687. Turning to the RSF, our student found that the desired parent compound (**10.8**) is 2,4-methano-2H-pentaleno[1,6-bc]furan (RF number 30679). He also was able to reconstruct the *CA* Index Name of **10.3** by using the numbering on the structural diagram of **10.8** in the RSF. The *CA* Index Name of **10.3** is 2,4-methano-2H-pentaleno[1,6-bc]furan, 6-carboxy-2-methyl-.

10.1.3. Notations

A line chemical notation is a description and representation of the structure of a chemical compound by a unique and unambiguous linear sequence of letters and numbers. It seems to be a superior indexing tool (computerized or manual) to chemical names or molecular formulae.

Among the various notation systems which have been developed during the years (e.g. Dyson, Hayward, Wiswesser), the Wiswesser Line Notation (WLN) is the best known and the most widely used. The WLN uses the smaller number of symbols than any other notation system—41 symbols only. The symbols used are the 26 upper-case (capital) letters (A–Z), the 10 arabic numerals (0–9—the zero is slashed when printed or written in order to distinguish it from the letter O), four punctuation marks (ampersand, &; hyphen, -; slash, /; asterisk, *), and a blank space. The blank space is used as a shift-key in order to double the number of available characters; thus characters which are preceded by a blank space have a totally different meaning to those which are not. The 41 characters used in the WLN are available on any typewriter or computer keyboard. No special modification of equipment is needed in order to be able to use the WLN (contrary to some other notation systems).

The notation pays special attention to the various characteristic features of chemical structures, features which determine the compound's special properties, by having the functional groups or the ring structure symbols as the starting point of the notation. Alternative starting points and the choice of alternative structure paths are governed by the alphabetical order of the symbols.

All of the atomic symbols are used without change of meaning except K, U, V, W, Y, Cl, and Br. The single letters (K,U,V,W,Y) are used more efficiently for other purposes (e.g. U—double bond; V—carbonyl group; Y—carbon atom attached to three atoms other than hydrogen or doubly bonded oxygen). Chlorine and bromine are designated by one letter (G for chlorine, E for bromine). The two-letter symbols of other elements are slightly modified: they have hyphens on each side in order to distinguish them from a pair of two adjacent single-letter symbols (e.g. -SI-represents a silicon atom while SI represents an iodine atom bound to a sulphur atom). There are 'new atomic symbols': KA for potassium, UR for uranium, VA for vanadium, WO for tungsten, and YT for yttrium. Hydrogen is usually implicit although it is used sometimes (H). Arabic numbers are used for unbranched internally saturated aliphatic chains (e.g. 1 represents CH_3- or $-CH_2-$, 3 represents $CH_3CH_2CH_2-$, $CH_3CH_2CH=$, $CH_3CH_2C\equiv$, $-CH_2CH_2CH_2-$, $-CH_2CH_2CH=$, etc.)

A detailed description of the WLN and its various rules and procedures can be found in E.G. Smith and P.A. Baker, *The Wiswesser Line Formula Chemical Notation (WLN)*, 3rd ed., Chemical Information Management, Cherry Hill, N.J., 1976.

Some representative examples of WLN are as follows:

Compound	WLN
HI	[H
$CH\equiv CH$	1UU1
$CH_3CH=CH_2$	2U1
$CH_3COCH_2CH_3$	2V1
$NH_2CH_2CH_2CH_2NH_2$	Z3Z
$(OOCCH_2CH_2COO)Ca$	OV2VO & $-$CA $-$
diphenylamine	RMR
m-nitrophenol	WNR CQ
p-aminophenol	ZR DQ
phenol	QR
compound **10.3**	T555 B5 C 2AB K IO JH&&&TS

10.1.4. Connection tables

A connection table is a detailed list of all atoms and bonds in the chemical structure as well as a description of the ways in which every atom is connected to the structure. It is a complete description of the topology of the chemical structure of a molecule in a way which is suitable for computer storage and/or processing. To simplify things, hydrogen atoms are not included. They are assumed to fill all 'free valences'; thus a connection of a carbon atom by only one single bond to any other atom implies the presence of a CH_3 group, a connection of a carbon atom by only one double bond to any other atom or by two single bonds to two other atoms implies the presence of a CH_2 group.

A connection table may be stored in a way in which any connection between two atoms is cited twice, as shown in Fig. 10.1 (a redundant connection table). A much more efficient storage is achieved by citing each connection only once, as shown in Fig. 10.2 (a non-redundant connection table).

204

NODE	ATOM	NODE/ATOM	NODE/ATOM	NODE/ATOM
1	C	7 CSE	2 CSE	.
2	C	4 CDE	3 CSE	1 CSE
3	N	6 CSE	5 CSE	2 CSE
4	O	2 CDE		
5	C	3 CSE		
6	C	3 CSE		
7	C	1 CSE	9 RN	8 RN
8	C	10 RN	7 RN	
9	C	11 RN	7 RN	
10	C	12 RN	8 RN	
11	C	12 RN	9 RN	
12	C	11 RN	10 RN	

CDE – CHAIN DOUBLE EXACT
CSE – CHAIN SINGLE EXACT
RN – RING NORMALIZE (AROMATIC)

Figure 10.1   A redundant connection table

10.1.5. CAS Chemical Registry System

The CAS Chemical Registry System was originally developed by the CAS in order to enable it to determine whether a chemical substance reported in the scientific literature had been indexed previously by the *CA* and, if so, under what name. A Registry Number (RN or CAS RN) is assigned sequentially to each new compound entering the Registry File. The RN contrary to a notation or a chemical name, does not give the searcher any information about the structure of the compound. It is simply a unique identification number for a specific chemical substance (just like personal identification by a Social Security Number or an ID Number).

Originally the CAS Registry System contained all the chemical substances which have been reported since the beginning of 1965. It included new substances which have been synthesized or identified since that time, and also known compounds for which new data have been reported since January 1965. Starting in 1984 the CAS began the registration of substances which were reported in the literature prior to 1965 and have not been mentioned again since then. Plans are to cover the period 1920–65, a project that would take a few years to complete.

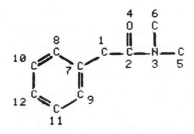

NODE	ATOM	NODE/ATOM	NODE/ATOM
1	C	2 CSE	
2	C	3 CSE	
3	N	6 CSE	5 CSE
4	O	2 CDE	
5	C		
6	C		
7	C	1 CSE	8 RN
8	C	10 RN	
9	C	7 RN	
10	C	12 RN	
11	C	9 RN	
12	C	11 RN	

CDE - CHAIN DOUBLE EXACT
CSE - CHAIN SINGLE EXACT
RN   - RING NORMALIZE (AROMATIC)

Figure 10.2   A non-redundant connection table

Only a small number of compounds which were registered were actively studied or produced for general use. About 76% of the substances in the CAS Chemical Registry System have been reported only once in the literature; on the other hand, 1% of the substances account for 45% of the literature citations.

The CAS RNs are used to identify and review chemical substances not only throughout the range of CAS publication and services but also in various other data banks and data bases (e.g. *Chemical Information System, Registry of Toxic Effects of Chemical Substances, BIOSIS, Index Medicus, International Pharmaceutical Abstracts*), primary journals (e.g. *Angewandte Chemie*), handbooks and dictionaries (e.g. *Handbook of Injectable Drugs, Heilbrons Dictionary of Organic Compounds, Merck Index*).

The CAS RN includes up to nine digits arranged in three groups—aaaabb-bb-b. The digits in the b positions are always present, while positions a may be completely filled (e.g. 100859-23-2 hexamethylbenzene biradical), completely empty (e.g. 70-25-7 *N*-nitrosoguanidine), or partly filled (e.g. 7782-

39-0 deuterium, 10028-17-8 tritium). The RNs are computer assigned in chronological order; the last digit is a computer-generated checking digit. Similar compounds, isomers, and isotopically labelled compounds usually have far-apart RNs (e.g. the RN of D-aspartic acid is 1783-96-6, L-aspartic acid 56-84-8, and the DL compound 677-45-8; the RN of dicyclohexylcarbodiimide or, as it is known by its *CA* Index Name, cyclohexaneamine, *N*,*N'*-methanetetraylbis-, is 538-75-0, whereas that of the fully tritiated compound is 60212-78-4). However, when similar compounds or isomers are reported for the first time in the same paper they may have consecutive RNs (e.g. the RNs of 2-undecyne, 3-undecyne, and 4-undecyne are 60212-29-5, 60212-30-8, and 60212-31-9, respectively. These three compounds were first reported in *J. Chromatogr.*, **119**, 451 (1976).

In general the larger the RN the newer is the compound. However, one has to remember that the year of registration does not necessarily correspond to the year in which the compound was first reported in the *CA*. A substance registered late one year may not be reported in the *CA* until early the next year.

Year	Registry Numbers
1975	53965-72-3—57866-57-6
1976	57866-58-7—61446-08-0
1977	61446-10-4—65229-05-2
1978	65229-06-3—68936-58-3
1979	68936-59-4—72467-41-5
1980	72467-42-6—76081-79-3
1981	76081-80-6—80386-99-0
1982	80386-00-6—84081-55-0
1983	84081-56-1—88343-57-1

During the years 1965–74 certain classes of substances received RNs from blocks of numbers set aside for each class (e.g. polymers were assigned numbers in the 9000-00-0 range). Since 1984, pre-1965 compounds are registered together with new compounds.

Information about the RNs of various chemical substances is available in the *CA* Chemical Substance Index (or *CA* Subject Index prior to 1972) and in the *CA* Formula Index. The CAS RN appears in square brackets following the *CA* Index Name. However, when one needs to assign a CAS RN to a substance one should turn to the *Registry Handbook*. The *Registry Handbook* is composed of two parts, the Number Section which is available only in printed form and the Common Name part which is available only in microform.

The Number Section of the *Registry Handbook* relates CAS RNs to the corresponding *CA* Index Name and the molecular formula of the compound. The *Registry Handbook* Number Section was published in 1971 in seven books covering the RNs which had been assigned during the years 1965–71. It is followed by an annual supplement covering the RNs assigned in each calendar year. A *Registry Handbook-Registry Number Update* is published annually and includes all the RN changes which took place in the whole Registry System from its beginning up to the last RN assigned in that year. Thus the last edition of the *Registry Handbook-Registry Number Update* gives a cumulative list of all RN changes which have occurred since 1965, and it should be used prior to using the Number Section of the *Registry Handbook* and its supplements. This update provides the user with the tool to determine whether a CAS RN is still valid. A CAS RN can be replaced by another number or it may no longer be in use. There are cases in which multiple registrations of the same substance occur, when different structural details are given in the original literature for the same substance by different authors, when more than one structure was postulated by various scientists, or when some error was made during the substance input process into the Registry System. When a multi-registration is recognized, a decision (based on careful review of the original literature) is made as to which RN should be retained and which one (or ones) should be discontinued. A cross-reference is then given from the discontinued RN to the active one. Sometimes a substance which was reported by name only is found later on not to be a unique chemical entry. In such cases the RN is removed from the Registry System and would not be included in any further CAS publication or service.

The second part of the *Registry Handbook* is the Common Name part which links about 600,000 common names to about 375,000 RNs and is composed of the Name Section and the Number Section. The Name Section links common substance names (arranged alphabetically) to their corresponding RNs and molecular formulae; the Number Section links RNs to the molecular formulae, *CA* Index Names, and common names.

There are online chemical dictionaries that link RNs, *CA* Index Names, common names, and some structural features (varied from dictionary to dictionary). Some of the dictionaries are: *CANON* on Questel, *CHEMDEX* on Orbit, *CHEMNAME* and *CHEMSIS* on Dialog, *CNZZ* on Data Star, *REG* File on STN International. RNs retrieved searching the various dictionaries can be transferred automatically (in the REG File) or by a special command (e.g. MAPRN on Dialog, PRINTSELECT on Orbit) to any other data base containing CAS RNs which are held by the vendor.

One should be aware that there are CAS RNs to compounds which have not been reported in the *CA*. These 'unavailable compounds' are compounds registered as components only (e.g. an acid known only in its salt form), ring parents (ring systems known only in their substituted form), compounds registered because of contracts of CAS with the American Federal Government Agencies, and compounds registered so recently that they have not yet been reported in the *CA*.

10.2. Substructural search

Before being able to search for the biological properties of compounds having rings **10.1** and **10.2** (see p. 196), one has to identify all known chemical compounds having such a structural feature. This is done by carrying out a substructural search. How is a thing like this done?

As we are dealing with a structure which is a ring or part of a ring system, one would expect that the *Ring System Handbook* would be a good search tool. Unfortunately it is not so, the *Ring System Handbook* has only two indexes: Ring Formula Index and Ring Name Index. It is interesting to know that its predecessor, the *Parent Compound Handbook* had seven indexes: Ring Analysis Index; Ring Substructure Index; Parent Name index; Wiswesser Line Notation Index; Parent Formula Index; Parent Registry Number Index; and Stereo Parent Index. Although the *Parent Compound Handbook* had been replaced in 1984 by the *Ring System Handbook* it could still be used, remembering that no new post-1983 ring systems are included in it.

The first stage in this search is identifying the various ring systems containing ring **10.1** or **10.2**; this is done with the aid of the Ring Substructure Index of the *Parent Compound Handbook*. The Ring Substructure Index connects the elemental ring analysis of each component of a ring system (except C_5 and C_6) to the *CA* Index Name and Parent Compound Identifier of the parent compound. It is arranged according to the compound ring formula. The compound ring formula is followed in parentheses by a component line formula which describes the arrangement of the elements in the ring skeleton. The various parent compounds are arranged according to their ring analysis.

In this search, the ring formula is S_3C_2 and the component line formula is S_3C_2. The following data were found under $S_3C_2(C_2S_3)$ or $C_2S_3(S_3C_2)$ (the Ring Substructure Index is a rotated one):

$C_2N_2-C_2S_3-C_5-C_5$
4,7-Methanobenzotrithiolo[5,6-*c*][1,2-]diazete [LWZDK] (**10.9**)

10.9

C_2S_3
1,2,3-Trithiolane [GVQPH] (**10.1**)
1,2,3-Trithiole [GVQRZ] (**10.2**)

$C_2S_3-C_4N_2-C_6$
1,2,3-Tritholo[4,5-*b*]quinoxaline [GVQSF] (**10.10**)

10.10

$C_2S_3-C_5-C_5$
4-7-Methanobenzotrithiole [HCJBT] (**10.11**)

10.11

$C_2S_3-C_5-C_5-C_5$
4,8-Methano-4H-indeno[5,6-d]trithide (HPLST) (**10.12**)

10.12

$C_2S_3-C_5-C_5-C_5-C_5-C_5$
4,10:5,9-Dimethano-4H-cyclopenta[6,7]naphtho[2,3-d]-1,2,3-trithiole
[LVJDL] (**10.13**)

10.13

$C_2S_3-C_5-C_6$
4H-Indeno[5,6-d]trithiole [HPLTY] (**10.14**)

10.14

$C_2S_3-C_6$
Benzotrithiole [FHJGK] (**10.15**)

10.15

$C_3N-C_2S_3-C_5-C_5$
4,7-Methanobenzotrithiole[5,6-b]azete [LWVBC] (**10.16**)

10.16

In order to identify the various compounds having the above parent structures (**10.1** and **10.2**; **10.9** to **10.16**) The Chemical Substance Indexes of *CA* Volumes 103–96, the 10th and 9th Collective Chemical Substance Index, and the 8th, 7th, and 6th Collective Subject Index were searched—47 different compounds were located being derived from the desired parent compounds. However, it was clear to our chemist that there are many compounds that were not retrieved: coordination compounds having the desired structure features—coordination compounds are indexed in the *CA* under the metal which is considered to be the parent compound; compounds which were reported prior to 1972 and were not reported in the literature since 1972—nomenclature changes occur at the beginning of 1972 (compound **10.17** was indexed under benzotrithiole,5-methyl-2-oxide [13688-81-2] in the 9th Collective Chemical Substance Index and under thiosulphurous acid (H_2S_3O), cyclic *S,S*-(4-methyl-*o*-phenylene) ester [13688-81-2] in the 8th Collective Subject Index); and compounds in which the ring system is perhaps not a parent compound but a substituent (e.g. benzenesulphonamide, *N*-(5,5-diphenyl-1-1,2,3-trithiolan-4-ylidene)-4-methyl-[65665-39-6] (**10.18**).

10. 17

10. 18

Is there a way to identify the various compounds online? One approach is to use an online chemical dictionary; Dialog Chemical Dictionary Files are shown below (as of 1 May 1986):

File 30: *CHEMSEARCH*	*CA* 100(26)−104(14)	
File 300: *CHEMZERO*	1965–Mar. 85	1,064,711 Sbus
File 301: *CHEMNAME*	1967–Dec. 85	1,547,123 Sbus
File 328: *CHEMSIS*	1967–1971	818,826 Sbus
File 329: *CHEMSIS*	1972–1976	1,115,864 Sbus
File 330: *CHEMSIS*	1977–1981	1,347,612 Sbus
File 331: *CHEMSIS*	1982–Sep. 85	929,622 Sbus

Search of the Formula of Ring (FR) field of six of the seven Dialog Files (*CHEMSEARCH* does not have an FR field) results in the identification of

104 compounds (as it turned out later on, 29 of those compounds are not reported in the *CA*). One can see that the effectiveness of the manual search is 62.6%. Strangely enough STN *REG* File does not have an FR field.

Theoretically a substructure search could be carried out in chemical dictionary files by string searching of chemical name fragments; in practice this results in low (sometimes very low) precision.

The best system for substructural search is a retrieval system based on connection tables. Queries are expressed in a structural form either alphanumerically or graphically. Those systems due to the explicitness of the connection tables yield complete recall and 100% precision.

Three substructural search systems are commercially available on host computers: CAS ONLINE substructural search system; Telesystems—DARC (Description, Acquisition, Retrieval, and Correlation); and CIS SANSS (Structure and Nomenclature Search System). Other systems have been developed for inhouse usage, e.g. CROSSBOW by ICI, COUSIN by the Upjohn Company. The search software of DARC and SANSS are commercially available for inhouse usage with private data bases. CAS ONLINE search software is not available for purchase for inhouse usage; however, company chemical compounds data bases can be mounted and searched as private data bases on the STN computers. Chemical structure search software packages are also developed and marked by various software houses. The most known package in this class is MACCS (Molecular Access System) which has been developed by MDL.

The whole Registry System File of the CAS is searchable using the CAS ONLINE *REG* File which at January 1st 1988 contains about 8,800,000 chemical substances. Few chemical substances data bases are searchable using the DARC system: *EURECAS*—the CAS Registry System (excluding polymers) from 1965; *POLYCAS*—polymers registered in the CAS Registry System since 1965; *SPECTRA–NBS* Mass Spectra File. The chemical compounds which are included in any of the CIS Files are searchable by SANSS. SANSS was the first substructure search system which became available to the public (mid 1970). Unfortunately, due to stoppage of US Government support (originally it was a government sponsored project) no improvement or upgrading has been carried out on the system since then. Indeed, it is the only system out of the three that does not have graphical input or output capabilities.

Using the CAS ONLINE *REG* File (STN International) for the substructure search of structures **10.1** and **10.2** results in the retrieval of 106 compounds. Two of the compounds were not retrieved on the Dialog chemical dictionaries search: 1,2,3-trithiole-4,5-dicarbonitrile, 2-oxide (**10.19**) and 4,8-methano-4*H*-indeno[5,6-*d*]trithiole, 3a,4a,5,7a,8,8a-hexahydro-, homopolymer (**10.20**). The first compound is a pre-1965 registration compound and as such is not included in the other vendors' chemical dictionaries. The second compound was reported in the *CA* 102(6):46326p which was published on 11 Feb. 1985; as such the compound should have been included in Dialog

212

File 331 (*CHEMSIS* 1982-Sep. 85) but it is not there. Close examination of Dialog File 30 (*CHEMSEARCH*) reveals the compound there; however, the file does not have an FR field and the compound could not have been retrieved in that way.

10.19

10.20

As of May 1987 Dialog File 30 (*CHEMSEARCH*) has been removed from the system. All other Dialog Chemical Dictionary Files (Files 300, 301, and 332) are now updated on a monthly basis, usually 3 to 4 weeks after the end of the month.

The substructure search on the *REG* File is shown below:

FILE 'REGISTRY' ENTERED AT 05:20:39 ON 25 APR 86
COPYRIGHT 1986 BY THE AMERICAN CHEMICAL SOCIETY

⇐ STRUCTURE ─────────────────────────────────**(1)**
ENTER NAME OF STRUCTURE TO BE RECALLED (NONE):
ENTER (DIS), GRA, NOD, BON OR ?:GRA R5, NOD 1 2 3 S, BON ALL S, DIS

```
            2
            S
1      .    .    3
    S.        . S
        .        .    ←────────────────────────────(2)
        .        .
        C:::::::::C
5                4
```

ENTER (DIS), GRA, NOD, BON OR ?:END ←────────────────**(3)**
L1   STRUCTURE CREATED

⇐ STRUCTURE
ENTER NAME OF STRUCTURE TO BE RECALLED (NONE):L1
ENTER (DIS), GRA, NOD, BON OR ?:BON 4-5 D, DIS

```
            2
            S
1      .    .    3
    S.        . S
        .        .    ←────────────────────────────(4)
        .        .
        C:::::::::C
5                4
```

ENTER (DIS), GRA, NOD, BON OR ?:END
L2   STRUCTURE CREATED

⇐ SEARCH ←——————————————————————————————(5)
ENTER LOGIC EXPRESSION OR QUERY NAME (END):L1 OR L2 ←———(6)
ENTER TYPE OF SEARCH: (SSS), FAMILY, OR EXACT:. ←—————(7)
ENTER SCOPE OF SEARCH: (SAMPLE), FULL, OR RANGE:. ←————(8)
SEARCH INITIATED (5:22:28) ←————————————————(9)
L3   ANSWER 1
     *
     *

SAMPLE EXECUTING

SEARCHED	SEARCH TIME	ANSWERS	PROJECTED	ANSWERS
0.072%	00.00.08	2	2 TO	8587

ENTER (DIS), STATUS, OR ?:STATUS ←————————————(10)

SAMPLE SEARCH COMPLETE

SCREENED	PASSED	ITERATED	MATCHED	INCOMPLETE	ANSWERS
392020	5	5	4	0	4
5.030%	0.011%	100.000%	80.000%	0.000%	

SEARCH TIME	DISPLAYED	PROJECTED ITERATIONS		PROJECTED	ANSWERS
00.00.28	1	5 TO	232	4 TO	198

ENTER (DIS), STATUS, OR ?:END ←———————————————(11)

SAMPLE SEARCH COMPLETE

FULL FILE PROJECTION:COMPLETE. ←———————————(12)
L3   4 SEA SSS SAM L1 OR L2

⇐ SEARCH ←————————————————————————————(13)
ENTER LOGIC EXPRESSION OR QUERY NAME (END):L1 OR L2
ENTER TYPE OF SEARCH: (SSS), FAMILY, OR EXACT:.
ENTER SCOPE OF SEARCH: (SAMPLE), FULL, OR RANGE:FULL ←———(14)
SEARCH INITIATED (5:24:40)
L4   ANSWER 1
     *
     *

SEARCH EXECUTING

SEARCHED	SEARCH TIME	ANSWERS	PROJECTED ANSWERS	
0.295%	00.00.06	1	1 TO	1352

ENTER (DIS), STATUS, OR ?:STATUS ←————————————(15)

SEARCH EXECUTING

SCREENED	PASSED	ITERATED	MATCHED	INCOMPLETE	ANSWERS
5088599	135	135	53	0	53
65.300%	0.002%	100.000%	39.259%	0.000%	

SEARCH	TIMEDISPLAYEDPROJECTED	ITERATIONSPROJECTEDANSWERS
00.03.08	1	153 TO 259 53 TO 113

ENTER (DIS), STATUS, OR ?:STATUS ←——————————————————————————(16)

FULL FILE SEARCH COMPLETE

SCREENED	PASSED	ITERATED	MATCHED	INCOMPLETE	ANSWERS
7792597	259	259	106	0	106
100.000%	0.003%	100.000%	40.926%	0.000%	

SEARCH	TIMEDISPLAYEDPROJECTED	ITERATIONSPROJECTEDANSWERS
00.05.53	1	259 TO 259 106 TO 106

ENTER (DIS), STATUS, OR ?:END

FULL FILE SEARCH COMPLETE
L4   106 SEA SSS FUL L1 OR L2 ←——————————————————————————(17)

(1) The STRUCTURE command is used for building a structure. The GRAph subcommand is used to specify the atoms arrangements in the structure (R5 specifies a five-member Ring). The NODe subcommand is used to specify other atoms than carbon in the structure (carbon atoms are the default) (1 2 3 S specifies that atoms 1, 2, and 3 should be Sulphur). The BOnd subcommand specifies the various BOnds in the structure (all S specifies that all bonds should be Single). The DISplay subcommand asked for structure displayment.

(2) The structure which was created is displayed.

(3) The system is instructed that the structure building ENDed.

(4) A second structure in which the BONd between atoms 4 and 5 is a Double bond was created and displayed. There is no need to define a Normalized (aromatic) bond between atoms 4 and 5 as it is included in both the other structures (L1 and L2) in the definition of the BONd as Single and Double.

(5) The SEARCH command is used for searching the REG File.

(6) The Logic expression of the search is defined by the Boolean operator OR (search for structure L1 OR L2).

(7) The default of the search system is SSS (SubStructural Search); the default is expressed by entering a dot (.).

(8) A search on a Sample of the REG File (the system default) is carried out. It is recommended to carry a sample search in order to make sure that the search does not exceed the system limits.

(9) After the search is initiated, the first answer used to be displayed automatically (in the SUB formate). Since September 1986 the first answer is not automatically displayed.

(10) The STATUS command is used to check the detailed status of the search.

(11) The END command is used to exit the SEARCH command.

(12) The prediction is that a FULL REG File search of the above query will run to completion.

(13) The SEARCH command is used again.

(14) This time a FULL search is carried out.

(15) The STATUS command is used to inform us that 5.088,599 compounds (which are 65.3% of the REG File) have been searched; the search took 3.08 min. up till now and results in 53 answers. The prediction is for 53 to 113 answers.

(16) The STATUS command is used again to inform us that the search is complete

and 7,792,597 compounds have been searched; the search lasted 5.53 min and results in 106 answers.

(17) The system informs us that an answer file (L4) which contains the 106 answers has been created.

A smaller number of compounds was obtained, 95 compounds, when the substructure was carried out on Telesystems-DARC using the EURECAS File.

DARC SYSTEM FOR SUBSTRUCTURAL SEARCH * 85-12 *

CMD (BA,ST,CN,SP,BI,INFO) ? BA
** BA **
ELAPSED TIME ON EURECAS : 0.67
BASE (MINICAS/EURECAS/POLYCAS/UPCAS/ISI-IC/SPECTRA) ? EURE
** EURECAS DATABASE : 7 098 879 STRUCTURES
CN   :35-00-7 TO 100482-86-8

CMD (BA,ST,CN,SP,BI,INFO) ? ST   ←————————————————(1)
** ST **
-ST-(QT,QG,RF,RE,AA) ? QU   ←————————————————(2)
** QU **
-QU-(CN,RF,CA,GR,BO,AT,FS,CH,VE) ? GR   ←——————————(3)
** GR **
? 1-2-3-4-5-1   ←————————————————————————(4)
? FI   ←————————————————————————————(5)
-QU-(CN,RF,CA,GR,BO,AT,FS,CH,VE) ? AT   ←——————————(6)
** AT **
? C   ←————————————————————————————(7)
? S 1,2,3   ←————————————————————————(8)
? FI
-QU-(CN,RF,CA,GR,BO,AT,FS,CH,VE) ? BO   ←——————————(9)
** BO **
? SI   ←————————————————————————————(10)
? Z 4-5   ←————————————————————————(11)
? SI 4-5
? DO 4-5
? AR 4-5
? FI
-QU-(CN,RF,CA,GR,BO,AT,FS,CH,VE) ? FS   ←——————————(12)
** FS **
? 4 1,2,3   ←————————————————————————(13)
? 2 4,5   ←————————————————————————(14)
? FI

216

```
-QU-(CN,RF,CA,GR,BO,AT,FS,CH,VE) ? FI
-ST-(QT,QG,RF,RE,AA) ? RE  ←──────────────────────────(15)
** RE **
CN S RANGE (Y/N) ? N  ←──────────────────────────(16)
RESULT    :  1367  ←──────────────────────────(17)
NEXT LIST    :  18441  ←──────────────────────────(18)
CONTINUE (Y/N) ? Y  ←──────────────────────────(19)
RESULT    :  791
NEXT LIST    :  18441
CONTINUE (Y/N) ? N  ←──────────────────────────(20)
RESULT    :  791
** RE **    791 ANSWER(S) **


-ST-(QT,QG,RF,RE,AA)    ? AA  ←──────────────────(21)
** AA **
   12 ANSWER(S) FOR    100 , ULTIMATE ESTIMATION    94
   47 ANSWER(S) FOR    200 , ULTIMATE ESTIMATION   185
   55 ANSWER(S) FOR    300 , ULTIMATE ESTIMATION   145
   58 ANSWER(S) FOR    400 , ULTIMATE ESTIMATION   114
   70 ANSWER(S) FOR    500 , ULTIMATE ESTIMATION   110
   82 ANSWER(S) FOR    600 , ULTIMATE ESTIMATION   108
   89 ANSWER(S) FOR    700 , ULTIMATE ESTIMATION   100
** AA **    95 ANSWER(S) **


-ST-(QT,QG,RF,RE,AA)    ? LN  ←──────────────────(22)
** LN **
** AA **    95 ANSWER(S) **  ←──────────────────(23)
         *
         *
         *

**FINS**
ELAPSED TIME ON EURECAS :    8.82
* DARC * IS OFF , BYE
09 H 43    *    86.04.27


CLR PAD (00) 00:00:10:29 383 46
```

(1) The ST command is used for structural search.
(2) The QU command is used for structure input.
(3) The GR command is used for graph input.
(4) The system is instructed to build five-member rings.
(5) The FI command is used to inform the system that the graph command ended.
(6) The AT command is used to specify the kind of AToms in the structure.
(7) All atoms are specified as Carbon atoms.
(8) Atoms 1,2,3 are specified as Sulphur atoms.
(9) The BO command is used to specify the kind of BOnds in the structure.

(10) All bonds are specified as SIngle bonds.

(11) The bond between atoms 4 and 5 is specified as a variable one: SIngle, DOuble, or ARomatic.

(12) The FS command is used to specify the maximum number of attachments allowed.

(13) Atoms 1,2,3 can have a maximum of four attachments.

(14) Atoms 4,5 can have a maximum of two attachments.

(15) The RE command is used for fragment search.

(16) The whole EURECAS File is searched.

(17) The first fragment file contains 1367 structures.

(18) The second fragment file contains 18441 structures.

(19) The system inquires whether to continue fragment searching; the answer is Yes.

(20) After obtaining the second fragment file the system is instructed to stop fragment searching.

(21) The AA command is used for atom by atom search, a search which results in 95 answers.

(22) The LN command is used to prepare a list of the compound numbers.

(23) There is a list containing 95 answers (or numbers).

Out of the 11 'missing' compounds, 10 are polymers which should be retrieved using the POLYCAS File. The other compound was compound **10.19**, which is missing in the Dialog dictionaries as well, a compound which is a pre-1965 registration compound.

Another substructure search locates all compounds having azacyclobutadiene (azete) moiety (**10.21**) in them. Turning to the *Parent Compound Handbook* one would find in the Ring Substructure Index that there are hundreds of entries under C_3N (in this case the compound ring formula and the component line formula are the same). Indeed, it is impractical even to try and run a manual search. The same picture occurs when an online search is carried in the Dialog chemical dictionary files; e.g. in File 328, *CHEMSIS* 1967-71, 4638 compounds were retrieved when the FR field was searched for NC_3. Examining a few of them gave structures such as **10.22** and **10.23** (no relevant answer was found in a sample of 20 compounds). However, a substructure search of the REG File of CAS ONLINE retrieved 89 compounds within 8 minutes.

10.21

10.22

10.23

FILE REGISTRY' ENTERED AT 05:36:24 ON 25 APR 86
COPYRIGHT 1986 BY THE AMERICAN CHEMICAL SOCIETY

⇐ STRUCTURE
ENTER NAME OF STRUCTURE TO BE RECALLED (NONE):.
ENTER (DIS), GRA, NOD, BON OR ?:GRA R4, NOD 1 N, BON ALL N, DIS

```
1                      2
   N+ + + + ++ C
   +            +
   +            +
   C+ + + + ++ C
4                      3
```

ENTER (DIS), GRA NOD, BON OR ?:END
L1   STRUCTURE CREATED

⇐ SEARCH
ENTER LOGIC EXPRESSION OR QUERY NAME (END):L1
ENTER TYPE OF SEARCH: (SSS), FAMILY, OR EXACT:.
ENTER SCOPE OF SEARCH: (SAMPLE), FULL, OR RANGF:.
SEARCH INITIATED (5:37:52)
L2   ANSWER 1
       *
       *

SEARCH EXECUTING -   1 ANSWERS

SEARCH EXECUTING

SEARCHED	SEARCH TIME	ANSWERS	PROJECTED ANSWERS
0.321%	00.00.12	1	1 TO 1591

ENTER (DIS), STATUS, OR ?:STATUS

SAMPLE SEARCH COMPLETE

SCREENED	PASSED	ITERATED	MATCHED	INCOMPLETE	ANSWERS
419660	33836	1000	1	0	1
5.385%	8.062%	2.955%	0.100%	0.000%	

SEARCH TIME	DISPLAYED	PROJECTED ITERATIONS	PROJECTED ANSWERS
00.01.27	1	618470 TO 638120	304 TO 952

ENTER (DIS), STATUS, OR ?:END

SAMPLE SEARCH COMPLETE

FULL FILE PROJECTION:INCOMPLETE. WILL EXCEED SYSTEM LIMITS.
←——(1)
L2   1 SEA SSS SAM L1

⇐ SCREEN   ←——(2)

ENTER SCREEN EXPRESSION OR (END):1852 OR 1853 ←————————(3)
L3   SCREEN CREATED

⇐ SEARCH
ENTER LOGIC EXPRESSION OR QUERY NAME (END):L1 AND L3 ←——(4)
ENTER TYPE OF SEARCH: (SSS), FAMILY, OR EXACT:.
ENTER SCOPE OF SEARCH: (SAMPLE), FULL, OR RANGE:.
SEARCH INITIATED (5:42:01)
ANSWER NOT YET AVAILABLE
SEARCH EXECUTING   0 ANSWERS

SEARCH EXECUTING
SEARCHED	SEARCH TIME	ANSWERS	PROJECTED ANSWERS	
0.323%	00.00.20	0	0 TO	O

ENTER (DIS), STATUS OR ?:STATUS

SAMPLE SEARCH COMPLETE
SCREENED	PASSED	ITERATED	MATCHED	INCOMPLETE	ANSWERS
434245	900	900	6	0	6
5.572%	0.207%	100.000%	0.666%	0.000%	

SEARCH TIMES	DISPLAYED	PROJECTED ITERATIONS		PROJECTED ANSWERS	
00.00.53	0	14537 TO	17763	6 TO	238

ENTER (DIS), STATUS, OR ?:END

SAMPLE SEARCH COMPLETE
FULL FILE PROJECTION:COMPLETE ←————————————————(5)
L4   6 SEA SSS SAM L1 AND L3

⇐ SEARCH
ENTER LOGIC EXPRESSION OR QUERY NAME (END):L1 AND L3
ENTER TYPE OF SEARCH: (SSS), FAMILY, OR EXACT:.
ENTER SCOPE OF SEARCH: (SAMPLE), FULL, OR RANGE:FULL
SEARCH INITIATED (5:44:33)
L5   ANSWER 1
          *
          *
SEARCH EXECUTING -   1 ANSWERS

SEARCH EXECUTING
SEARCHED	SEARCH TIME	ANSWERS	PROJECTED ANSWERS	
2.965%	00.00.17	1	1 TO	134

ENTER (DIS), STATUS, OR ?:STATUS

SEARCH EXECUTING
SCREENED	PASSED	ITERATED	MATCHED	INCOMPLETE	ANSWERS
4735237	11903	11903	52	0	52
60.765%	0.251%	100.000%	0.436%	0.000%	

220

SEARCH TIMES	DISPLAYED	PROJECTED ITERATIONS	PROJECTED ANSWERS
00.02.49	1	19051 TO 20125	52 TO 120

ENTER (DIS), STATUS, OR ?:STATUS

FULL FILE SEARCH COMPLETE

SCREENED	PASSED	ITERATED	MATCHED	INCOMPLETE	ANSWERS
7792597	18993	18993	89	0	89
100.000%	0.243%	100.000%	0.468%	0.000%	

SEARCH TIMES	DISPLAYED	PROJECTED ITERATIONS	PROJECTED ANSWERS
00.05.16	1	18993 TO 18993	89 TO 89

ENTER (DIS), STATUS, OR ?:END

FULL FILE SEARCH COMPLETE
L5   89 SEA SSS FUL L1 AND L3 ←————————————————————(6)

(1) The prediction is that a FULL REG File search would not run to completion. This is caused by the fact that the instructions to the system were to find all substances containing three or more carbon atoms, one or more nitrogen atoms, and one or more rings. Those are commonly occurring features which are found at least in about 10% of the compounds in the REG File.

(2) The SCREEN command is used in order to select manually screens for the search.

(3) The four-member ring is defined by screens 1852 and 1853. Screen 1852 represents four-membered rings with no points of fusion to other rings (DDDD), screen 1853 represents four-membered rings with one to four points of rings fusion (DDDT, DDTT, DTDT, DTTT, or TTTT).

(4) The logic expression of the search is defined by combining the structure L1 with the screen L3 using the Boolean operator.

(5) The prediction is that a FULL REG File search of the above query will run to completion.

(6) The system informs that an answer file (L5) which contains 89 answers has been created.

CHAPTER 11

Environmental Impact and Control, Occupational Hygiene, Safety, Toxicity

The amount of information and regulation concerning the toxicity and/or the environmental impact and control of various chemical compounds as well as the number of safety and occupational hygiene regulations are growing at an exponential rate, a rate which is at least comparable to (if not exceeding) the 'knowledge explosion' rate. Those topics should concern every chemist even if he is not directly involved with handling specific chemicals or running chemical reactions and/or chemical processes in the laboratory, the pilot plant, or the industrial plant.

In this daily work every chemist is involved with at least one of the following subjects which are related to safety and similar topics:

(a) General safety practice in the laboratory, the pilot plant, or the industrial plant.
(b) Dangerous physical characteristics of various chemical compounds and mixtures (e.g. explosive character, thermal stability, flammability).
(c) Hazardous chemical reactions and side-reactions. The best known example is the formation of 2,3,7,8-tetrachlorodibenzo-p-dioxin (**8.1**) as a by-product in the industrial production of 2,4,6-trichlorophenol from 1,2,4,6-tetrachlorobenzene (Reaction 11.1). When the reaction temperature exceeds 160 °C, **8.1** becomes a major product. This was the reason for the 'Seveso Accident' which happened on 10 July 1976, an accident which not only threatened the health of the people living nearby but also affected the social and the economic life of the whole area.
(d) Short- and long-term effects of chemicals on human beings, test animals (especially when data on human beings are unavailable), vegetation, and aquatic life (e.g. the effect of thalidomide on pregnant women).
(e) Problems concerning the transportation, handling, storage, and discharge of dangerous materials (e.g. poisons, radioactive waste). The best known case is the 'Bhopal Disaster' which took place in the Indian city

222

on 2 December 1984 and in which many tons of the highly volatile and highly toxic methyl isocyanate (11.1) burst from a storage tank.
(f) Pollution control.
(g) Local, national, and international regulations concerning safety and related topics.
(h) Biodegradability.

(11.1)

CH$_3$NCO

11.1

One should keep in mind that standards, regulations, and environmental/safety/toxicity ramifications change rapidly. A substance whose usage was considered safe yesterday may be forbidden today (owing to its toxicity, mutagenicity, or other hazard properties). Examples are the famous cyclamate controversy in the USA, the discovery that certain chemicals which were used extensively in the laboratory and/or industry are carcinogenic or mutagenic [e.g. chloromethyl methyl ether (11.2) which was used as a chloromethylating agent, N-nitroso methyl urea (11.3) which was used as a source for the formation of diazomethane, and various laboratory and industrial solvents such as benzene and chlorinated hydrocarbons).

ClCH$_2$OCH$_3$

11.2

NH$_2$CON⟨CH$_3$ / NO

11.3

11.1. The safety literature

There is a large number of journals dealing with the various aspects of environmental impact and control, occupational hygiene, safety, toxicity, and related topics. Similarly to any other scientific paper, the papers which are published in these journals are abstracted and indexed by the various abstract services.

The average chemist today is pressed so hard with his own scientific work and with his updating reading programme that he cannot and should not include any of the safety literature in his updating programme. However, he should be aware of the existence and the scope of the various secondary sources of the safety literature and use them whenever he is in need of information on safety and related topics. One should remember that many of the laboratory accidents could have been avoided or could have been ended with only minimal injuries and damage if the practising chemists had been aware of the various properties of the chemicals involved. The original scientific literature in this area should be screened and read by those chemists who are involved in one way or another in safety enforcement in their institutions (e.g. safety officers) as well as by the environmental chemists.

11.1.1. Primary literature

There are many journals devoted to general or specific aspects of occupational hygiene, pollution, safety, toxicity, and similar topics. One may find the relevant journals in Ulrich's *International Periodical Directory* under the following section heads: Environmental Studies, Industrial Health and Safety, Water Resources. There are general scientific journals which include safety information and reports. The data may appear as a 'Letter to the Editor' (e.g. *Chemistry in Britain*, *Chemical and Engineering News*), as a news story (*Chemtec*, *Chemical Age*) or as letters, reports, and/or articles (*Journal of Chemical Education*).

11.1.2. Secondary sources

The secondary sources of safety and related topics are abstracts services and journals, handbooks, compendia, and data banks.

The various abstracts services and journals include well known general chemical and non-chemical abstracts services which have been discussed before as well as services and journals dealing mainly with one or more of the aspects of safety and related topics.

The various safety, toxicity, and similar topics are covered by the *CA* mainly in six sections: Sections 1, 4, 17, 59, 60, and 61 covering Pharmacology, Toxicology, Food and Food Chemistry, Air Pollution and Industrial Hygiene, Waste Treatment and Disposal, and Water, respectively. More specific coverage dealing with safety in the chemical and nuclear industry, as well as the health and safety of workers in those areas or workers with hazardous materials, is given in the *CA Selects — Chemical Hazards, Health, and Safety*. The coverage of this *CA Selects* also includes the effects of human exposure to hazardous substances and the hazardous properties of chemical compounds and reactions. Publications dealing with animal toxicology, carcinogens, environmental pollution, mutagens, or teratogens are usually not covered by this *CA Selects*, unless human safety or health is emphasized in them. Such publications are covered by other *CA Selects — Carcinogens, Mutagens and Teratogens*; *Drugs and Cosmetic Toxicology*; *Environmental Pollution*; and *Food Toxicology*. Information and reports about specific incidents such as chemical plant explosions, fires, chemical spills, and leakages could be located in *Chemical Industry Notes*.

Ascatopics also gives a good coverage of the subject; the topic — *Chemical Hazards — Health, and Safety* has a similar coverage to the *CA Selects — Chemical Hazards, Health, and Safety*. Other topics which cover similar subjects are: *Air Pollution — Effects, Source, and Control*; *Chemical Residues in Food Beverages*; *Environmental Mutagens*; *Heavy Metals in the Environment*; *Marine Pollution*; *Water Pollution — Effects, Source, and Control*; and *Chemical Carcinogens*.

Biological Abstracts covers the bioligical and/or chemical pollution and toxicology, biological safety, and various environmental aspects. A good coverage is found in the following *BIOSIS Standard Topics*: *Air Pollution*; *Monitoring of Environmental Pollution*; *Pesticide Residues in Food*; *Pesticide Residues in Soil*; and *Sewage Disposal and Sanitary Measures*. BIOSIS also published some speciality abstracts journals; one of interest is *Abstracts on Health Effects on Environmental Pollution* (*HEEP*). HEEP deals mainly with chemical substances (excluding medicinal chemicals) that affect or are potentially harmful to human health as well as substances that have potentially harmful effects on vertebrates and/or invertebrates. It covers the adverse effects on humans as well as their possible effects on wildlife.

Abstracts of US Government-sponsored research and Governmental Agency reports on environmental topics such as air, health, noise, pesticides, radiation safety, solid wastes, and water are reported in the NTIS bulletin *Environmental Pollution and Control*.

A much wider survey of health and pollution control is carried out by the *Environmental Health and Pollution Control*. This abstracts journal is a monthly publication of *Excerpta Medica* and its coverage includes over 15,000 scientific journals and serials.

Air, land, freshwater, and marine pollution, environmental actions, noise, radiation, sewage and waste water treatment, toxicology and health, and waste management are topics covered by *Pollution Abstracts*. It is a monthly abstracts publication covering about 2500 different sources — journals, serials, books, and conference proceedings.

Environmental Abstracts is another abstracts service; it abstracts documents dealing with air pollution, chemical and biological contamination, energy, environmental topics such as design, ecology, education, and international aspects, food and drugs, land use and misuse, non-renewable and renewable resources, ocean and water pollution, population control, radiation, solid wastes, transportation, and wildlife. It is a monthly publication of the Environmental Information Center (EIC), covering about 5000 different sources (journals, serials, conferences proceedings, governmental and institutional reports, national newspaper stories).

A current awareness journal dealing with laboratory safety measures is *Laboratory Hazards Bulletin*. A similar journal dealing with the chemical industry safety measures is *Chemical Hazards in Industry*. Both journals are monthly publications of the Royal Society of Chemistry (RSC). They report and abstract articles dealing with safety measures and potential hazards as well as legislation affecting the well-being of chemists working in the chemical laboratory and/or in the chemical industry. Their online version is available via DATA STAR, Pergamon Infoline and STN.

Safety science in its broad definition is covered by the *CIS Abstracts* and/or by *Safety Science Abstracts*. *CIS Abstracts* is published by the International Labor Organization through its Centre Internationale d'Information de Securité et d'Hygiène du Travail (CIS). It covers over 1500 journals dealing with all aspects of informational industrial safety (it is available online via ESA–IRS and Telesystems Questel). *Safety Science Abstracts* covers the various aspects of safety—safety to the scientist, to humans, and to the environment, phenomena that threaten safety, technology that establishes this safety, and research and application in this area. It is published monthly by the Cambridge Scientific Abstracts and covers over 17,000 publications (it is available online via Pergamon Infoline).

Most of the journals cited in all these abstracts services are not chemistry journals and they are usually not available in a chemistry library. If and when a relevant abstract is found, one would have to obtain the original article using the help of an outside source.

11.2. General safety practice

Various governmental and non-governmental agencies cover various general safety practices in the chemical laboratory, the pilot plants, or the industrial plants. One should be familiar with the various regulations (which usually differ from country to country) dealing with the different safety aspects.

Among the agencies devoted mainly to safety in the USA is the US National Institute for Occupational Safety and Health (NIOSH), which serves as a clearinghouse for information regarding occupational safety and health. The results of NIOSH-supported research are disseminated through the NTIS. The NIOSH produces the Occupational Safety and Health (NIOSHTIC) data base which covers all aspects of the occupational safety and health fields (e.g. behavioural sciences, chemistry and biochemistry, control technology, hazardous wastes, safety, toxicology). It covers about 400 journals and 70,000 monographs and technical reports. In the UK there is the Royal Society for the Prevention of Accidents (ROSPA) which deals mainly with industrial safety and safety regulations. It disseminates information via its publication programme (e.g. *Occupational Safety and Health, ROSPA Bulletin*) and by other means such as conferences and exhibitions. Another UK organization is the British Safety Council (BSC) which produces among other things the journals *Safety* and *Safety and Rescue*.

Many other organizations devote part of their activities to occupational safety and health. Among them are the various national atomic energy commissions or authorities (e.g. nuclear waste treatment), the various Standards

Institutes (e.g. standards of various safety equipments), the American Chemical Society (Division of Chemical Health and Safety), and the MCA (which issues *Safety Data Sheets* and *Accidents— Case History Guide for Safety in the Chemical Laboratory*).

11.3. Safety in the chemical laboratory

The prevention of accidents, injuries, fires, and other losses in the chemical laboratory should be a prime interest of chemists involved in laboratory work. The various aspects of safety in the laboratory include chemical storage requirements and systems, codes and standards on safety and health, electrical safety in the laboratory, emergency procedures, eyes, face, skin, and respiratory protection, flammable hazards and control, hazard identification and control, health hazards of chemicals, laboratory waste disposal, and safe handling of compressed gases. One should be aware of the sources covering the various safety topics and use them whenever there is a need. One should remember that all those sources are designed to be consulted and used *before* a hazardous situation arises.

A general source of laboratory information is the second edition of the *CRC Handbook of Laboratory Safety*, edited by N.V. Steere and published in 1971 by the CRC Press Inc. This is a compendium of over 70 topics related to laboratory safety, followed by Tables of Chemical Hazard Information. Other comprehensive sources of information are the 1984 publication of the Royal Institute of Chemistry entitled *Health and Safety in the Chemical Laboratory*; and N.T. Freeman and J. Whiteman, *Introduction to Safety in the Chemical Laboratory*, published by Academic Press, New York, 1983.

Data on hazard chemicals can also be found in special compilations such as N.I. Sax *et al.*, *Dangerous Properties of Industrial Materials*, 6th ed., published by Van Nostrand Reinhold, New York, 1984 (covering over 6000 chemical substances) and L. Bretherick, *Handbook of Reactive Chemical Hazards*, 3rd ed., published by Butterworths, London, 1985 (covering over 4000 compounds). Whereas Sax's book deals mainly with the toxicological hazards of various chemicals, Bretherick's covers the fire and explosion hazards of specific chemicals as well as the fire and explosion hazards in carrying out specific chemical reactions; hence those two books are complementary to each other. A pocket version of Sax's book entitled *Rapid Guide to Hazardous Chemicals in the Workplace* has been published recently. Other sources of this type are *The Sigma-Aldrich Library of Chemical Safety Data* and H.H. Fawcett's book *Hazardous and Toxic Materials: Safety, Handling and Disposal* published by Wiley, New York, 1984. The Sigma-Aldrich compilation covers over 7000 chemical compounds giving the compound name, its CAS RN, chemical structure, physical properties; toxicity and regulatory data; health hazards, first aid statements; handling, storage, spill, and fire extinguishing instructions; decomposition products; recommended waste-disposal methods. Fawcett's book is more of a kind of primer on the subject. The book covers the following topics: laboratory safety (followed by an excellent bibliography), toxicity, hazards, fires, explosions, protective equipments.

Hazard and toxicity information on chemical compounds can be found in the fifth edition of *Heilbron's Dictionary of Organic Compounds* (see p. 84). Information about the toxicity and hazard effects of many chemicals is given, followed by key references for more detailed information.

The *Journal of Chemical Education* has a column on laboratory safety entitled 'Safety in the Chemical Laboratory' which has been edited by N.V. Steere. Articles published in the years 1967–74 have been reproduced in a series of three paperbacks entitled *Safety in the Chemical Laboratory* edited by N.V. Steere. A fourth volume covering articles published during 1975–80 in the *Journal of Chemical Education* as well as any other safety material published in the journal during the above period was edited by M.M. Renfrew. Those four paperbacks were published by the Division of Chemical Education of the American Chemical Society.

In 1981 the National Academy of Sciences/National Research Council (USA) published a report entitled *Prudent Practice of Handling Hazardous Chemicals in Laboratories*. It deals with the various laboratory hazards in sufficient depth to give the reader a very good overview of the constitution of a good effective safety programme in the chemical laboratory.

11.4. Conducting a Safety-related Search

A chemist interested in obtaining data on environmental impact and control, occupational hygiene, safety and/or toxicology should consult the safety department in his organization (it could have other names such as environmental and hygiene, loss prevention, etc.). Some of the required data may already be available and handled immediately, while the rest will be searched for by professionals. In many institutions and in some industries, owing to their small size or for some other reason, the chemist himself would have to search for pertinent data.

When conducting a search in this area the following steps should be followed:

(a) Contact the manufacturer as a first source in order to obtain the latest safety information about a specific chemical. However, if the manufacturer is located abroad this may take too long and another source of information should be used.
(b) Consult a basic reference book or a basic reference collection.
(c) Some of the safety information, especially that related to hazardous chemicals, hazardous chemical reactions, or laboratory accidents, are reported as 'Letters to the Editor' in chemical news journals such as *Chemical and Engineering News*, *Chemistry in Britain*, and *Chemistry and Industry*. These letters are abstracted in the *CA*, so one can locate

them by searching the *CA*. More information concerning those specific hazards and/or accidents could be obtained by searching the *SCI* using the corresponding 'Letter to the Editor' as a starter. Information about the hazardous effects of a specific chemical could be obtained using the *SCI*, employing the original article reporting the usage of the compound as a starting point.

(d) If you cannot locate any safety information on a specific compound, try to locate data on analogous or homologous compounds. Such data may sometimes, but not always, provide useful information; in any case, if you are making assumptions based on such data, consult a safety expert or a toxicologist.

(e) If you are watching for trends and developments in specific topics, pay special attention to news, information, and data originating in countries such as Canada, Japan, Sweden, and/or West Germany, which are known for their special concern for the environment and its control (e.g. the potential hazards of mercury were first identified in Japan).

Those who are involved directly in obtaining data on the various aspects of safety and hygiene should be aware of the following articles:

P.A. Carson and K. Jones, 'Effective Management of Safety and Hygiene Information to Promote Safe Handling of Chemicals', *Journal of Hazardous Materials*, **9**, 305–314 (1984).

R.B. Wright, 'Recent Advances in Information-Storage and Retrieval Relevant to Occupational Hygiene', *Annals of Occupational Hygiene*, **24**, 313–324 (1981).

11.4.1. Dangerous physical and biological characteristics of chemical compounds

People working with chemicals should use appropriate safety procedures if those chemicals are known to be toxic, irritating to the skin or eyes, mutagenic and/or carcinogenic, spontaneously flammable upon heating or contact with air or water, or spontaneously explosive upon heating or shock. However, how can one know in advance about the hazards of various chemicals one is working with?

All chemicals could be divided into two main classes—the 'old' or 'known compounds' and the 'new compounds'. The 'known compounds' are those which are commonly used as starting materials, reagents, or solvents in chemical reactions or are produced by the chemical and pharmaceutical industries (e.g. food additives, insecticides, drugs, paints). The various properties of those compounds are usually well known so one can find well in advance if and which safety procedure should be used when handling them. 'New compounds' are usually reaction products which have been synthesized for the first time or new chemical reagents which have been recently introduced. It is not rare to find that dangerous properties of 'new compounds' are reported a short time after their introduction as chemical reagents.

Tetrabutylammonium permanganate (11.4) was reported at the beginning of 1978 as an oxidizing agent in organic synthesis [T. Sala and M.V. Sargent, *Chem. Commun.*, 253 (1978)]. Shortly afterwards the spontaneous ignition of the compound was reported [J.A. Morris and D.C. Mills, *Chem. Ind. (London)*, 446 (1978); *Chem. Br.*, **14**, 326 (1978)]. The same is true with a similar oxidizing agent—benzyltriethylammonium permanganate (11.5) whose properties were reported in the literature at the beginning of 1979 [H.J. Schmidt and H.J. Schafer, *Angew. Chem. Int. Ed.*, **18**, 68, 69 (1979)] only to be followed shortly afterwards by information about its thermal and impact instability [H. Jager, J. Lutolf and M.W. Meyer, *Angew. Chem. Int. Ed.* **18**, 786 (1979]; H.J. Schmidt and H.J. Schafer, *Angew. Chem. Int. Ed.*, **18**, 787 (1979)]. If Schmidt and Schafer were aware of the spontaneous ignition of the tetrabutyl salt (11.4) we are sure they would have examined and reexamined the benzyltriethyl salt (11.5) before making the claim that it is thermally stable at 100 °C and is resistant to impact. Indeed, they realized in their second work that although **11.5** was stable at 100 °C for 5 min it ignited at this temperature after 7 min at 90 °C after 25 min and at 80 °C after 90 min of heating.

$$(C_4H_9)_4NMnO_4 \qquad\qquad C_7H_7(C_2H_5)_3NMnO_4$$

11.4 **11.5**

It has been said earlier that many laboratory accidents could have been avoided or at least caused much less damage once the practising chemists were aware of the properties of the chemicals they were dealing with. Let us examine one case of a reported accident—unfortunately one out of many.

The explosion of 2,2-bis(dihydroperoxy)propane during a class demonstration is recorded in the literature [*Chem. Eng. News*, **62**(37), 4 (1984)]. It is hard to believe that an organic peroxide was used for a demonstration without any apparent knowledge of the danger involved in its usage.

(a) It has been reported that the 2,2-bis(dihydroperoxy) propane (11.6) used was a solid. The above compound is reported in the literature as a liquid [N.A. Milas and A. Golubovic, *J. Am. Chem. Soc.*, **1**, 6461 (1959)].

$$CH_3-\underset{\underset{OOH}{|}}{\overset{\overset{OOH}{|}}{C}}-CH_3$$

11.6

(b) Two other compounds also result from the reaction of acetone and hydrogen peroxide, the linear dimeric acetone peroxide (11.7) and the cyclic trimeric acetone peroxide (11.8).

230

$$(CH_3)_2 \underset{\underset{H}{\overset{O}{|}}\ \underset{H}{\overset{O}{|}}}{COOC(CH_3)_2}$$

11.7

11.8

(c) From the description in the literature it looks as though the demonstrator was using the cyclic trimer or a mixture of the cyclic trimer and linear dimer, but definitely not the monomeric compound.

(d) There are few reports in the literature about the explosive properties of the dimeric and/or trimeric acetone peroxide (**11.7**; **11.8**) e.g. G.A. Olah, N. Yoneda, and D.G. Parker, *J. Am. Chem. Soc.*, **99**, 483–488 (1977); A. Noponen, *Chem. Eng. News*, **55**(8), 5 (1977); A.D. Brewer, *Chem. Br.*, **11**, 335 (1975); H. Seidl, *Angew. Chem. Int. Ed.*, **3**, 640 (1964); W. Treibs, *Angew. Chem. Int. Ed.*, **3**, 802 (1964); E.G.E. Hawkins, *Organic Peroxides: Their Formation and Reactions Span*, London, 1961, p. 140.

(e) It is very clear it was only a miracle that the explosion did not cause much more damage.

Let us consider an organometallic chemist who is interested in repeating a reaction which was reported by A.J. Blakeney and J.A. Gladysz in *Inorg. Chim. Acta*, **53**, L25 (1981) (Reaction 11.2).

$$K^+[(CO)_4 FeSi(CH_3)_3]^- + CH_3OSO_2F$$

$$\longrightarrow (CO)_4 FeSi(CH_3)_4$$

$$(11.2)$$

In order to carry out the reaction our chemist had to order the methylating reagent—methyl fluorosulphonate. As he could not locate the desired compound in the latest Aldrich and Fluka catalogues which were on his desk he decided to carry out an online search of the *Fine Chemicals Directory*.

FINE CHEMICALS DIRECTORY — VERSION 2
Copyright 1983, Fraser-Williams (Scientific Systems) Ltd.
— Updated April 1985)

Enter your request
/ S MF = CH3FO3S

SET   1: 2   MF = CH3FO3S
/ DISPLAY 1 F9/1-2

Item 1
Accession No: 37039
Mol. Formula: CH3FO3S

No data for number 37039 (item 1)

Item 2
Accession No: 36289
Mol. Formula: CH3FO3S

WLN & Suffix: WSFO1

Supplier	Cat. No.	Compound Name
ALFA	20834	METHYL FLUOROSULFONATE 95%
COLUMBIA	M-2522	METHYL FLUOROSULFONATE/MAGIC METHYL 97%
FLROCHEM	F05095	METHYL FLUOROSULPHONATE (*WITHDRAWN*)
K & K	K28143	METHYL FLUORO SULFONATE (*WITHDRAWN*)
PARISH	2287	METHYL FLUOROSULFONATE
PCR INC	P11212-8	METHYL FLUOROSULPHONATE
RIEDEL	61395	METHYL FLUOROSULFONATE PROSYNTH

Examining the 1985 catalogue of Riedel-de-Haen our chemist found that the desired compound is a corrosive one (a fact that did not surprise him) and put an order for 20 g of the material.

What our chemist did not realize was that methyl fluorosulphonate is an extremely toxic as well as carcinogenic material. How could he find this information? By searching the properties of the desired compound.

Indeed, while nothing is reported about the compound in the *Laboratory Hazard Bulletin* our chemist could find eight relevant reports by carrying out an online search of the *CA*.

(1) D.M.W. Van den Ham and D. Van der Meer, *Chem. Eng. News*, **54**(36), 5 (1976).
(2) D.M.W. Van den Ham and D. Van der Meer, *Chem. Ind. (London)*, 782 (1976).
(3) J. Ashby, D. Anderson, and J.A. Styles, *Mutat. Res.*, **51**, 285 (1978).
(4) R.W. Alder, M.L. Sinnott, M.C. Whiting, and D.A. Evans, *Chem. Eng. News*, **56**(37), 56 (1978).
(5) R.W. Alder, M.L. Sinnott, M.C. Whiting, and D.A. Evans, *Chem. Br.*, **14**, 324 (1978).
(6) M. Hite, W. Rinehart, W. Braun, and H. Peck, *Am. Ind. Hyg. Assoc. J.*, **40**, 600 (1979).
(7) J.E. Cumming and J.F. King, *Chem. Br.*, **15**, 329 (1979).
(8) D. Bartholomew and I.T. Kay, *Tetrahedron Lett.*, 2827 (1979).

The first two references are two identical letters describing the death of a Dutch chemist who had worked with the compound. The third reference reports a positive Ames test of methyl fluorosulphonate. The next two references are identical letters discussing the potential hazard of the compound. Toxicological data of the compound is given in reference 6. The next reference is a letter to the editor discussing the potential carcinogenicity of the methyl ester. The last reference is a synthetic work in which methyl fluorosulphonate was used as a methylating agent. There is a footnote in the paper which was copied into the *CA* abstract—'MeO₃SF is a highly toxic reagent, great care must be exercised in its use'.

A manual search of the 8th Collective Subjects Index, the 9th–10th Collective Chemical Substances Indexes, and the Chemical Substances Index of Volumes 96–103 of the *CA* retrieved eight relevant references (it is interesting to note that while the online search took about 2 min at a cost of about $3.00, the manual search lasted 95 min). One reference that was retrieved in the online search (reference 8) was not retrieved in the manual search; however, the same letter that was reported in references 1 and 2 was also published in *Chemistry in Britain* [D.M.W. Van den Ham and D. Van der Meer, *Chem. Br.*, **12**, 326 (1976)] and was retrieved manually.

Another source that could be used is the *Occupational Safety and Health*—the data base of the NIOSH. Indeed, an online search of the data base retrieved two answers—references 3 and 6 [*Mutat. Res.*, **51**, 285 (1978); *Am. Ind. Hyg. Assoc. J.*, **40**, 600 (1979)].

A manual or online search of RTECS gave all the toxicological data related to the compound; the results of the online search are shown below.

RTECS Search System (Version 5.1/11.0 April, 1985) ($55/hr)
Toxicity data in RTECS have not been critically evaluated

Option? TSHOW 421-20-5

CAS RN 421-20-5   NIOSH number LP0720000

FLUOROSULFURIC ACID, METHYL ESTER
C-H3-F-O3-S

SKIN AND EYE IRRITATION DATA:
eye-rbt 100 mg/4sec rns SEV   AIHAAP 40,600,79

MUTAGEN DATA:
mmo-sat 4 ug/plate   MUREAV 51,285,78

TOXICITY DATA:
ihl-rat LC50:5 ppm/1H   AIHAAP 40,600,79

orl-mus LD50:112 mg/kg   AIHAAP 40,600,79
skn-rbt LDLo:455 mg/kg   AIHAAP 40,600,79

DATA REFERENCES:
MUTAGEN
SKIN AND EYE IRRITANT
EPA GENETIC TOXICOLOGY PROGRAM, JANUARY 1984
MEETS CRITERIA FOR PROPOSED OSHA MEDICAL RECORDS RULE
FEREAC 47, 30420,82

As a matter of fact, if our chemist was going to use *Heilbron's Dictionary of Organic Compounds* in order to study some of the physical properties of the methyl fluorosulphonate he would have found a toxicity tagging attached to the compound (compound M-01821) followed by the sentence 'Toxic vapours which cause pulmonary oedema' and a RTECS number—LP 0720000. He would thus have immediately become aware of the toxicity of the compound.

The methyl fluorosulphonate case should teach us a few things:

(a) The importance of checking safety data of chemical substances.
(b) Different data bases cover different aspects of the same subject. There is a different emphasis among the various data bases which results in obtaining varied search results depending upon the data base used. An example is a safety search concerning methylisocyanate (**11.1**) (Table 11.1).

Table 11.1   Coverage of the toxicity of methylisocyanate (**11.1**) by various safety data bases

Data base	Coverage period	No. of hits
Chemical Hazards in Industry	1.84–7.85	79
Laboratory Hazards Bulletin	7.81–8.85	2
NIOSH	1.73–7.85	10

(c) When an online search is carried out one should remember that only part of the literature could be searched in that way. Online search usually goes back 15–20 years (sometimes only 5–10 years), and whenever an online search gives unsatisfactory or negative results a manual search covering the pre-online coverage period should be carried out. Indeed no data were located about fluoromethyl sulphonate using *Chemical Laboratory Hazards*; however, one should remember that its coverage started only in 1981, all the material retrieved from the *CA* was pre-1980.

234

(d) Abstracting and indexing are the results of information analysis carried out by information analysts having a scientific background. This analysis depends upon the analyst and varies from analyst to analyst, as could be seen from the three different abstracts and indexing terms in the *CA* to the identical letters of Van den Ham and Van der Meer which were published in *Chemical and Engineering News*, *Chemistry in Britain* and *Chemistry and Industry*, respectively.

Abstracts:

(1) A fatal laboratory accident involving methyl fluorosulphate [421-20-5] vapours was reported. The victim experienced initial coughing, but appeared to recover. However, 6 h after vapour inhalation a pulmonary oedema rapidly developed which proved fatal.
(2) Accidental inhalation of $MeOSO_2F$ [421-20-5] vapour following a breakage caused death by pulmonary oedema.
(3) Accidental inhalation of a small amount of $MeOSO_2F$ vapour, following a flask breakage, caused a fatal pulmonary oedema.

Keywords:

(1) methyl fluorosulphate poisoning; safety methyl fluorosulphate
(2) fluorosulphate safety, oedema lung methyl fluorosulphate
(3) fluorosulphate safety methyl; methyl fluorosulphate inhalation death

Index terms:

(1) Poisoning—methyl fluorosulphate
    Safety—of methyl fluorosulphate
    421-20-5—poisoning by
(2) Health hazard—from Me fluorosulphate
    Death—from methyl fluorosulphate inhalation
    Lung, disease or disorder—oedema, from methyl fluorosulphate
    421-20-5—death from inhalation of
(3) Health hazard—from Me fluorosulphate
    Edema—of lung, from Me fluorosulphate
    Lung, disease or disorder—oedema, from Me fluorosulphate
    421-20-5—health hazard from inhalation of, lung oedema in relation to

(1) *Chem. Eng. News*, **54**(36), 5 (1976)— *CA* 85:138198
(2) *Chem. Br.*, **12**, 362 (1976)— *CA* 86:126403
(3) *Chem. Ind. (London)*, 782 (1976)— *CA* 87:10798

CHAPTER 12

The Patent Literature

The patent literature is a major information source describing all developments of new technologies. One has to keep in mind that only 10–20% of the data reported in the patent literature is being published or cited in other information sources (journals, books, etc.). Even these 10–20% are disclosed in patents many years before their publication in any other media (e.g. patents were granted to the Ziegler-Natta polymerization catalysts 10 years prior to their first report in a non-patent publication). Patents are different from any other information sources (they are individual documents that stand on their own, they provide the basis for claims for an exclusive right to practise the invention(s) described therein, they contain also 'speculative' or 'general' information, etc.) and as such require a separate treatment.

The importance of the patent literature in one's current awareness programme has been discussed earlier (see p. 59). Patents yield, in addition to technological and scientific information, valuable data about the short- and long-range research interests of the various patent holders (corporation and/or inventors), and may help in predicting various corporate policies and strategies. They give information about the patentability and validity of new inventions, help in finding gaps where one can try to invent, or give one an idea for improving an invention. Although such improvements are patentable, they do not give one the right to use the invention, but it usually results in one getting much better licensing terms from the original patent holder.

A patent is a document disclosing to the public an invention of some kind (e.g. a new composition of matter such as a new catalyst or a new polymer; a new product such as a new solar energy collector or a new oven for making pizza; a new process such as a new way of manufacturing nuclear fuel). The inventor gives up the secrecy of his invention, getting in return a monopoly for a specific period (one has to remember that this monopoly does not automatically permit him to practise his invention; e.g. one has to get a permit from the corresponding health authorities for the production and sale of new drugs). It is true that the inventor is giving something up when he discloses his idea to the public, and somebody may be able to develop his idea further. However, without the patent, somebody else might come up with the same idea, patent it, and prevent the original inventor from using it. Further, without the protection of a patent, the inventor may lose his invention as it may be disclosed through 'leaks' or dishonesty.

Patent laws (which will not be discussed in this chapter) differ from country to country, but there are four general legal requirements in the various countries which must be fulfilled in order that a patent be obtained—novelty, invention, utility, and completeness of disclosure.

Novelty: The invention must be new. The invention should not have been patented anywhere previously, nor the idea for it published anywhere prior to making the patent application. In order to prove priority of the novelty of the idea, the inventor (prospective patentee) should keep reliable and adequate dated records of his work.

Invention: The patent should include an inventive step. It should be clear that the subject to be patented would not have been obvious, in view of the state of the art, at the time the invention was made.

Utility: The item must be useful at the time of the invention. New chemical compounds which are not recognized as useful for anything cannot be patented, scientific theories cannot be patented, etc.

Completeness of disclosure: The patent description must be adequate to enable a person skilled in the art to make and/or use the item.

One has to remember that the interpretation of these requirements may differ from country to country.

A patent is valid only in the country in which it was issued, which is the reason why the same invention is patented in several countries. These patent duplications and multiplications (which are known as patent equivalents and patent families), as well as the increasing number of inventions followed by the time-consuming examination processes, lead to a large backlog in the various Patent Offices. A few approaches have been introduced during the last 10–20 years in order to catch up with the backlong and to shorten the time-consuming examination procedures. One approach is to publish unexamined patent applications, thus drastically shortening the publication time (such is the case in France, Japan, and West Germany, amongst other countries, which are known as rapid patent-publishing countries). One should distinguish between an issued patent and a published unexamined patent application, as it is well known that not all applications end up as patents—some of them are withdrawn by the inventors while others are rejected by the patent examiners on various grounds. Another approach is to combine efforts when a patent for the same invention is sought in more than one country by having one submission, one search, and even one examination [e.g. the Patent Cooperation Treaty (PCT)]; however, the patent has to be registered in each country separately. The latest approach is the multi-national patent—e.g. the European Patent System and the European Economic Community Patent System.

Dates are very important in the life of a patent; the most important date is the Priority Date which is also known as the Convention Date. It is the earliest filing date of the patent in any one of the member countries of the International Convention for the Protection of Industrial Properties (The Paris Convention). A patent application filed in any Convention country within 12 months of an earlier filing in any other Convention country has all the legal priorities of the first filing date. One has to keep in mind that sometimes there is usage of partial or multiple priorities. This arises when two or more related applications filed in one Convention country are the basis of a single application in another country. This is caused because of the different patent laws in various countries.

The patent document of the various countries are similar—they all consist of a cover page, specification and drawing (if any), followed by the claims. The cover page carries a standard set of information. The specification is a description of the invention. It discloses in full, clear, concise, and exact language what is new in the invention and its advantage over what is already known. This part also includes specific experimental procedures–examples. The examples could be compared to the experimental part of a journal, and one should be able to repeat them as easily as repeating a standard experimental procedure. The last part is the claims section, which is the expression and representation of the legal aspects of the patent, the claiming of what the inventor sees as his invention or discovery. The tendency is for the claims to be as broad as possible.

Information on topics such as new developments in patent classification, patent searching (manual and online), technological assessments, and forecasting in the patent field and similar ones could be found in *World Patent Information*. It is a quarterely journal published by Pergamon Press in co-operation with the WIPO and the Commission of European Communities.

12.1.   Chemical patents

There are six patent types: composition of matter, processes, machines, manufactures, plants, microorganisms and animals, designs. All six types of patents have larger or smaller chemical interest but the first four are the most important ones from the chemist's viewpoint.

Most, but not all, major countries now grant patents for chemical compounds per se. Other countries (such as most East European and Latin American countries, Austria, Greece, Spain) allow only claims to processes for the production of chemical compounds. There are some countries that have special patent provisions for pharmaceutical, foodstuff, agrochemical, and/or insecticides, as well as for intermediates for any of these compounds.

Patents are granted for new chemical compounds of known structure. In the case of unknown structure the compound could be patentable once it is defined in terms of its properties or in terms of how it was made. However, all patentable compounds should have an industrial application—be useful. Compounds may be new and useful but not patentable once their properties are obvious (e.g. a new compound obtained by replacing a bromine atom by chlorine). A compound which serves as an intermediate in the synthesis of a useful end product is considered as having industrial application and as such it is patentable.

A patent is granted not for one compound but for a group of compounds having some structural features in common as well as the same end usage. The claims in chemical patents contain claims for preparation, properties, and/or usage of substances with general non-specific structure(s) which are known as generic or Markush compound(s). The first such claim was made by E.A Markush in his patent 'Pyrazolone Dyes', US Patent 1,506,316 filed on 26 Aug. 1924. '...The process for the manufacture of dyes which comprises coupling with a halogen substituted pyrazolone, a diazolized unsulfonated material selected from the group consisting of aniline, homologues of aniline and halogen-substituted products on aniline...'

Patents which are of interest for chemists are not only patents on new compounds but also patents on new processes (new synthetic routes), new usage and application of well known chemical compounds (once the usage and/or application is not obvious), new compositions and mixtures of known compounds which possess industrial applications, as well as new machines and/or tools to be used in the chemical industry.

A good coverage of the various legal aspects as well as historical and political aspects of patents in general and of chemical patents in particular is given in P.W. Grubb's book *Patents for Chemists*, Clarendon Press, Oxford, 1982.

12.2. Searching the patent literature

There are various reasons and therefore there are as many approaches for searching the patent literature. There are searches which could and should be terminated as soon as answers to the problem in question are found; others would need all of the pertinent answers. All patent searches could be classified as follows: current awareness search, general (state of the art) search, patentability (novelty) search, infringement search, validity search, equivalent or family search, business-oriented search.

The current awareness search has been discussed earlier in Section 4.4.4.

The general search is designed to provide the chemist with general background, generally in a broad subject area. It is usually a search of the general literature of which the patent literature is a part.

The patentability search (also known as a novelty search) is designed to confirm that a specific product or process has never before been described or suggested in any way in prior art (reference published before filing a patent application). There are few problems running such a search: (a) how to design a search for a nonexistent reference, (b) how to recognize a reference describing the invention in a different vocabulary or context, (c) when and how to know that the search is over when no relevant references are found. The search strategy is similar to the one used in the case of general search;

however, its scope is wider in order to obtain related references. The search is carried out in more than one data base (usually three or four) and also covers newspapers, magazines, catalogues, etc. Special attention should be given to flash drops (answers satisfying the requirements of the search strategy but do not describe the invention of interest); they might be the nearest related prior art.

The infringement search is carried out in order to determine if unexpired patents have claims that might be infringed by a contemplated new commercial venture (process, product, etc.). The search is time limited—it concerns only patents that are still in force. It is limited also to patents of the country in which the venture in question would take place.

The validity search is very similar to the patentability search, but is carried out in greater depth and in a wider scope. One performs a validity search once dominating patents are found in an infringement search. The search is very comprehensive including the whole world patent literature as well as the scientific literature, newspapers, magazines, and monographs. Many times the most important sources are those that are uncommon and/or hard to find. One should try and read the original documents as the pertinent information may be some comment hidden in the text.

The equivalent or family search is carried out in order to locate equivalents or families of patents (patents granted in different countries for the same invention having the priority date of the first application). Identification of patent equivalents does not only eliminate the need of translation from a foreign language but also gives information about intentions and plans of the patent holder (usually an organization would not file a patent application in a foreign country unless it is planning to practise its invention there).

The business-oriented searches are carried out in order to get information regarding one or more of the above topics: licensing versus self-commercialization, company acquisitions, developing new business and/or markets, identifications of companies and/or personal working in a specific area, identification of new products, new processes, new materials, etc.

One usually uses secondary sources while searching the patent literature. These secondary sources could be general information sources whose coverage also includes patents (e.g. *CA*, *Physics Briefs*) or they may be sources that cover only the patent literature (e.g. *INPADOC Patent Gazette*, Derwent *WPI* and *CPI*, the various *Patents Newsletters* published by the Communication Publishing Group, Inc.). The various sources could be divided into a few groups: all technology patents—*INPADOC Patent Gazette*, Derwent *WPI*; chemical-related patents covering the whole area of chemistry—*CA*, Derwent *CPI*; specialized chemical-related patents—petroleum refining petrochemical-related patents are covered by *API Patent/Abstracts*, petroleum exploration and production-related patents are covered by *Petroleum Abstracts* (published by the Information Service Division of the University of Tulsa), paper-chemistry-related patents are covered by the *Abstract Bulletin* (published by the Institute of Paper Chemistry Appelton, Wis.), fibre-optic-related patents are covered by *Fiber Optics Patents Newsletter* (published by Communication Publishing Group Inc.) and by *Fiber Optics Patents* (published by Information Gatekeeper), etc. As a rule one should usually use the more general sources and not the specialized ones. The coverage of the general patent sources are much wider, their timelag is short (usually very short), and they are going to be published for a long time (e.g. *Fiber Optic Patents* mentioned earlier started publication in 1980 and ceased publication in 1985).

One should keep in mind that patent data bases of various countries are available for public online searching via various vendors; e.g. French patents are available on Telesystémes-Questel as *IMPI-1*, German patents are available on STN as *PATDPA* (plans are in progress to extend the data base to include graphics), USA patents are available on Dialog as *Claims* Files and on STN as *IFI* Files (a data base which is produced by IFI/Plenum), and European patents are available on Telesystémes-Questel as *EDOC*.

12.2.1. Chemical Abstracts

CA originally covered all chemical and chemical engineering patents which were issued to nationals and non-nationals of four countries only (France, Great Britain, Germany, and the USA). Over the years *CA* added to its coverage patents which were issued by various countries to their nationals. The assumption behind this policy was that all patents would eventually be documented in the country in which the individual or corporate patent applicant resides. When various Patent Offices started to publish unexamined patent applications, *CA* decided to change this policy and to cover all patents issued by certain countries and to extend the coverage to include unexamined applications, as it was clear that these documents represent the first description of a new invention — an invention that could have been developed in a country with a conservative patent policy. Today, the *CA* covers all chemical and chemical engineering patents issued by 18 countries and by two international organizations as well as selected patents issued by 8 other countries (patents issued to their residents only) (see Table 12.1).

CA is a general information source which covers many types of documents, patents being one of them. Patent abstracts are placed into one of the 80 sections of the *CA* according to their subject content; they are placed as a group at the end of the section. There is no specialized patent index and indexing is limited. Prior to 1979 *CA* indexed only chemical substances which were mentioned in the patents' examples. An expanding indexing policy of chemical substances started in January 1979 in which explicit compounds mentioned in the claims section were also indexed. The expansion was a gradual one, starting with patents reported in the Organic Chemistry Sections (Sections 21–34) and slowly extending into other sections. Inventors and patent holders are indexed in the Author Index. All Index entries related to patent documents are designated in the various indexes (Author

Table 12.1   *CA* patent coverage by country

Australia *	Israel *
Austria *	Japan *
Belgium *	Netherlands *
Brazil *	Norway †
Canada *	Poland †
Czechoslovakia †	Romania *
Denmark †	South Africa *
England *	Spain †
Finland †	Sweden †
France *	Switzerland *
Germany (East) *	USA *
Germany (West) *	USSR *
Hungary †	European Patent *
India *	PCT International *

* All patents of chemical or chemical engineering interest
(patents issued to national or non-national).
† Patents of chemical or chemical engineering interest issued to residents only.

Index, Chemical Substances Index, and General Subject Index) by the letter
P prior to the abstract number. There is at present one single Patent Index;
this index replaced the Numerical Patent Index and the Patent Concordance
at the beginning of the 10th Collective Period (1977).

The *CA* Patent Index lists all patent documents processed by the CAS (remember that patent equivalents are processed but not abstracted). It includes
entries for all newly abstracted documents, cross-references to the first abstracted patent document whenever the document processed by the CAS
deals with or is related to an invention that was described and abstracted
before in the *CA*. It also lists under the first abstracted patent documents all
the patent documents which are related to that particular invention (listing
of the whole patent family). The index is arranged alphabetically according to the country of issue code (a two-letter code, e.g. European Patent
Organization, EP; Israel, IL; Japan, JP).

One of the limitations of online search of the *CA* is that the Patent Index
(or the Patent Concordance prior to 1977) is not availble online to the
public. Searching for patents using a patent number is limited only to patents
documents which were abstracted by the *CA*. Even if one knows the number
of all the members of the patent family, how would one learn which one of
them was abstracted by the *CA*?

At present the CAS is building a separate extended patent data base, planning to market it some time during 1988.

12.2.2. INPADOC

The use of the weekly *INPADOC Patent Gazette* in one's current awareness
programme has been discussed earlier. This publication which contains the

bibliographical data on all the patent documents (from 49 countries and 2 international organizations; Table 12.2) which have been added to the INPADOC data base during the preceding week is complemented by a set of cumulated indexes which should be used for retrospective search.

Table 12.2 INPADOC patent coverage by country

Argentina	Korea, Republic of
Australia	Luxemburg
Austria	Malawi
Belgium	Mexico
Brazil	Monaco
Bulgaria	Mongolia
Canada	Netherlands
Cuba	New Zealand
Cyprus	Norway
Czechoslovakia	Philippines
Denmark	Poland
Egypt	Portugal
England	Romania
Finland	South Africa
France	Spain
Germany (East)	Sweden
Germany (West)	Switzerland
Greece	Turkey
Hong Kong	USA
Hungary	USSR
India	Yugoslavia
Ireland	Zambia
Israel	Zimbabwe
Italy	
Japan	European Patent
Kenya	PCT International

The *Patent Family Service* is an index which combines the equivalent patents on the basis of a common priority claim, thus giving the searcher a list of the whole collection of patent documents concerned with one invention. It is a monthly publication which accumulates monthly during the year. There is also a yearly cumulation which is published at the beginning of the year. It cumulates over the years up to a period of 5 years. Another index is the *Patent Classification Service* (arranged according to the IPC numbers). The IPC is a patent classification system maintained under the auspices of the WIPO, each patent usually having more than one IPC number. The system is built in five levels—the section level (all patents are divided into eight sections—A-H, of which C is chemistry and metallurgy), the class level (e.g. CO7 is organic chemistry), the subclass level (e.g. CO7B is organic chemistry, general methods and apparatus), the main group level (e.g. CO7B1/00 is organic chemistry, general methods and apparatus for hydrogenation), and the subgroup level whenever there is need for it. Other indexes are the

Patent Application Service (arranged according to the applicant's name), the *Patent Inventor Service* (arranged according to the inventor's name), and the *Numerical Data Base* (arranged according to the country and patent number). All of these indexes are published every three months, each issue being a cumulation of all the data collected since the beginning of the year. The January issue is a yearly cumulation which accumulates yearly up to a period of five years.

The great advantage of using the *INPADOC Patent Gazette* and its indexes is that it is the most comprehensive collection of the worldwide patent literature. Furthermore, it builds up and reports the various Patent Family Collections (since 1977 the INPADOC Patent Family Collections have been reported in the *CA* Patent Index). However, when using the *INPADOC Patent Gazette* one has to remember that it is only a bibliographic source; thus a patent title (as descriptive as it may be) in many cases would not give the desired information and the searcher would need more data in order to be able to decide about the relevance of the patent in question. One should also remember that the coverage period of INPADOC goes back only to the late 1960s and early 1970s (in some cases even to the early 1980s), depending on the country in which the patents were granted.

The INPADOC data base is also available online (from Orbit, Dialog, INKADATA). Paper and microfilm copies of patents reported in the *INPADOC Patent Gazette* can be obtained from INPADOC (either directly or through agents, e.g. IFI/Plenum in the USA). Other special services such as translation of Japanese patents, etc., are also available.

12.2.3. Derwent Publications

Derwent began its activity as an organization devoted to the rapid abstracting of patent documents at the early 1950s. In 1951 it started publication of Belgium patents as at the time Belgium was one of the most rapid patent-granting countries (16–18 months from application date). During the years the number of countries covered was expanded gradually. However, at that early time no attempts were made to index the chemical compounds claimed or exemplified in the various patents covered by the organization.

In 1963 Derwent began to cover patent specifications from 15 different countries covering all areas of pharmaceutical research, a service known as *Farmdoc*. Chemical structure could be retrieved from *Farmdoc* using a fragmentation code developed by Derwent. A fragmentation code is a series of codes (which may be letters and/or members) assigned to particular features of a chemical compound. The Derwent fragmentation code was originally based on the 960 position punch card (80 column, 12 lines) in which each position had a unique identifier. Subscribers received punch cards issued by Derwent on which bibliographic data as well as abstracts were printed, thus enabling them to run various Boolean logic searches (simple or complex ones) using a punch card sorter. In 1965 the fragmentation coding system was applied to all chemicals reported in agricultural patents (*Agdoc*). In 1966 a fragmentation code was introduced (based on a system developed by ICI) to cover all polymeric compounds reported in the various patent documents (*Plasdoc*). In 1970 the fragmentation system was expanded to cover all general chemical patents (*Chemdoc*)—the Derwent *Central Patent Index*—which in 1986 changed its name to the *Chemical Patent Index* (CPI). By the mid-1970s when the Derwent data base became available online, the fragmentation codes were no longer tied to 960; however, the Derwent chemical fragmentation code was completely revised in 1981 and there is no longer any ties to the punch card.

The CPI is divided into 12 classes (see Table 12.3) of which the first three are *Plasdoc*, *Farmdoc*, and *Agdoc*, respectively. Derwent covers chemical as well as non-chemical patents (e.g. human necessities, performing operations, transport, construction, mechanical engineering, instrumentation, computers, communication, electronic components, electric power) granted by 29 countries and 2 international organizations (see Table 12.4). Most of the Derwent publications have been discussed in Section 4.4.4.1. (Derwent patent publications). One additional Derwent Index publication is the *Electrical Patent Index* (*EPI*).

Table 12.3   Main classification of Derwent *CPI*

Classification	Code
PLASDOC, polymers	A
FARMDOC, pharmaceuticals	B
AGDOC, agricultural chemicals	C
Food, fermentation, detergents, cosmetics, etc.	D
CHEMADOC, general chemicals	E
Textiles, paper, cellulose	F
Printing, coating, photographic chemicals	G
Petroleum	H
Chemical engineering	J
Nuclear, explosives, protection	K
Glass, refractories, ceramics, electrochemistry	L
Metallurgy	M

Retrospective search of the data base could be carried out manually using the *WPI Gazette* Indexes which are available as COM. There are five indexes: Patentee Index, IPC Index, Accession Number Index, Patent Number Index, and Priority Index. Information concerning the various patents can be located in the corresponding *WPA* booklets, and copies of the original patent documents can be obtained for a nominal fee from Derwent. The various indexes are weekly indexes with quarterly (weeks 1–13, 1–26, and 1–39) and yearly cumulations. The current cumulation cycle is 10 years, starting in 1979. There is also a five-year cumulation for the years 1974–78 and a pre-1974 cumulation for chemical patents only.

Table 12.4 Derwent patent coverage by country

Australia	(maj)	Japan	(maj)
Austria	(min)	Luxemburg	(min)
Belgium	(maj)	Netherlands	(maj)
Brazil	(min)	Norway	(min)
Canada	(maj)	Portugal	(min)
China	(min)	Romania	(min)
Czechoslovakia	(min)	South Africa	(maj)
Denmark	(min)	South Korea	(min)
England	(maj)	Spain	(min)
Finland	(min)	Sweden	(maj)
France	(maj)	Switzerland	(maj)
Germany (East)	(maj)	USA	(maj)
Germany (West)	(maj)	USSR	(maj)
Hungary	(min)		
Israel	(min)	European Patent	(maj)
Italy	(min)	PCT International	(maj)

Major (maj) countries are countries whose patents are fully covered by Derwent; minor (min) countries are those covered only by title and bibliographic data.

The whole Derwent data base is also searchable online; however, contrary to most online files where once a record is added to the file it remains there unchanged, the record in the *WPI* changes. To the basic record may be added during the years one, two, or more equivalents, which may carry with them additional or new IPC numbers as well as new patent assignees.

Recently Derwent launched a statistical analysis software package— *PATSTAT Plus*. This software allows the user to analyse data downloaded from any of the WPI online hosts (Dialog, Orbit, Telesystem). It contains built-in graphics to display results in graphs and pie charts as well as being able to produce output in the form of spreadsheet tables.

12.3. Searching for a patent family

A patent family is a collection of patents covering the same invention issued at different times in different countries and having the same priority date. There is some difference in the resulting patents of the family due to variation in the patent laws in the various countries. As we have seen, most data bases index patents from a limited number of countries. In most cases only the first patent in a patent family is abstracted and/or indexed in depth. This patent is known as the 'basic' patent—it may or may not be the patent on which priority is based.

How can one locate an English translation of German patent number 2208542 dated October 1972? One should look for an equivalent or a family member patent that was issued in an English-speaking country (e.g. England, USA). The easiest way would be to use the *CA* 9th Collective Patent Concordance. Indeed, under DE 2208542 one finds the *CA* abstract number of the German patent (78 : 16453w) as well as three family members Brit

246

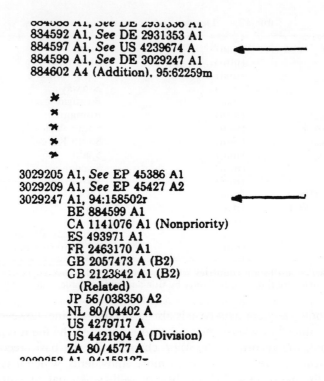

884566 A1, See DE 2931353 A1
834592 A1, *See* DE 2931353 A1
884597 A1, *See* US 4239674 A
884599 A1, *See* DE 3029247 A1
884602 A4 (Addition). 95:62259m

3029205 A1, *See* EP 45386 A1
3029209 A1, *See* EP 45427 A2
3029247 A1, 94:158502r
   BE 884599 A1
   CA 1141076 A1 (Nonpriority)
   ES 493971 A1
   FR 2463170 A1
   GB 2057473 A (B2)
   GB 2123842 A1 (B2)
   (Related)
   JP 56/038350 A2
   NL 80/04402 A
   US 4279717 A
   US 4421904 A (Division)
   ZA 80/4577 A
3029252 A1 94:158127r

Figure 12.1   Locating BE 884599 patent equivalents using the *CA* Patent Index

```
UV-hardenable silicone coating compsn. - contg. fluid
   polydialkyl-alkylepoxysilicone copolymer intermediate and bis-(aryl)
   iodonium salt catalyst, for imparting release
Patent Assignee: (GENE ) GENERAL ELECTRIC CO
Patent Family:
   CC  Number    Kind    Date     Week
   BE  884599     A      801201    8050   (Basic)
   NL  8004402    A      810205    8108
   DE  3029247    A      810219    8109
   NO  8002322    A      810302    8113
   SE  8005521    A      810309    8113
   GB  2057473    A      810401    8114
   FI  8002329    A      810331    8117
   FR  2463170    A      810327    8120
   JP  56038350    A     810413    8122
   US  4279717    A      810721    8132
   NO  8102470    A      810928    8143
   ZA  8004577    A      811117    8209
   CA  1141076    A      830208    8310
   US  4421904    A      831220    8402
   GB  2123842    A      840208    8406
   GB  2123842    B      840801    8431
Priority Data (CC,No,Date): US 63648 (790803);   CA 365247 (801121);   US
   272827 (810612)
Applications (CC,No,Date): GB 8025250 (800801);   GB 836003 (830304)
```

Figure 12.2   Locating BE 884599 patent equivalents using Derwent *WPA*

1388962, FR 2144199, US 3809689. All that now remains is to order the British or the American patent.

How can one build the patent family of which the Belgium patent 884599 is a member? Manually one can try and use the *CA* Patent Index which is based on the *INPADOC Patent Family Service*. Doing so one gets a list of family members (Fig. 12.1). One can obtain the same data online using INPADOC or Derwent data bases, searching for a patent family using a priority application number or a patent number. The INPADOC patent family search is relatively expensive as there is a surcharge of $45 for generating the patent family (a search for an English, French, or German language equivalent patent is much cheaper and it retrieves only a single family member, thus providing an access to a translation at reasonable cost). There is no surcharge for patent family search online using *WPI* (Derwent). However, one should remember that INPADOC covers nearly twice the number of patent issuing countries and organizations than Derwent. An online search for the patent family of the Belgium patent using the Derwent data base is shown in Fig. 12.2.

Not all patent families are simple families—sometimes two or more patent applications are combined for filing in another country, resulting in a family with more than one priority date. There are cases in which broadly defined inventions registered as one patent in one country have to be divided into two or more patents in another country having the same priority date. A more complicated family is one which includes amongst its members modifications and continuations; this happens when a patent application needs modifications after filing.

12.4.   Generic or Markush search

The retrieval of specific structures from a structural or compounds data base has been discussed in Chapter 10—Structural and Substructural Searches. Furthermore, the DARC and the CAS online have been enhanced to allow the retrieval of compounds through generic queries. However, these generic queries entered by the user are converted into a defined number of specific compound queries of the whole possible combination of the variable group. Those systems, although they are called 'generic', are in fact just an extension of the specific compound search system. None of the compound data bases have an option for the creation of a Markush data base; one should consider that one or two Markush claims can cover well over a few million compounds.

Markush search could be run using fragmentation codes, in a fragmentation coding system. In such a system one can represent all possibilities of the generic representations in terms of generic as well as specific groups claimed, disclosed, or exemplified. All that one has to do is to encode the various groups in such a way that they can be linked back together to the parent structure. Indeed generic groups and/or structures are searchable using fragmentation codes. However, in such a search the desired structures and/or compounds are followed by a large number of non-relevant answers.

At present the best known Markush search system is the one based on the Derwent chemical fragmentation codes. The Derwent chemical fragmentation codes have been revised a few times during the years in order to make search results more specific by providing additional molecular descriptions. Each revision (or improvement) increased the search precision. The search itself is carried out using the various chronological sections of the data base which results in a complex search retrieval, especially when searching for a complex generic structure. Complexity could be minimized by subdividing complex generic structures and searching for several simpler compounds, ignoring fragments whose coding is uncertain, omitting a Markush group when several functional groups are present in such a group, and omitting coding for carbon chain lengths and/or positions of ring substitutions (these are two of the trickiest parts of the coding system). This procedure simplifies the retrieval strategy very much while adding only a few false drops and reducing the opportunity for introducing errors, etc.

A much easier and more exact search system for Markush structures would be a graphics or a topological search system. It records the exact environment of every atom in the molecule and therefore can eliminate all the false drops obtained using the fragmentation coding system. Search input could be much easier and simpler by using graphic input. At the present time, Derwent in cooperation with Telesystèmes-Questel are working towards the creation of a Markush Darc system of a graphic generic data base search using the Darc software. Plans call for two steps: (a) single compound search against generic compound data base, (b) generic compound search against generic compound data base. Derwent started at the beginning of 1986 to index new compound specification of Sections B (pharmaceutical), C (agricultural), and E (general chemical) of the CPI into the graphic generic data base. Plans are to extend the coverage to other sections of the CPI as well as to known compounds reported in all chemical patents during the years 1978–88. Work on the production of a topological to fragmentation code conversion programme has been completed recently, a programme which is known as TOPFRAG. The idea behind this programme is to enable the compilation of fragmentation codes of a generic structure from its topological representation (its connection table), thus enabling an automatic regeneration of a search strategy for a Markush search (using fragmentation codes) on one or more of Derwent online hosts (Dialog, Questel, or Orbit).

CHAPTER 13

Chemical Marketing, Processes and Engineering Information

An industrial chemist needs information about reaction processes, chemical marketing, and other economic business data related to his particular problem. The industrial chemist (whether he is working in the laboratory, the pilot plant, or the industrial plant; whether he is involved in research and development, industrial processes, production, or carrying out executive duties) is part of the business world and as such should be aware of the various problems concerning the production, manufacturing, and/or marketing of his organization's products.

One has to bear in mind that there are no clear boundaries between technological and/or scientific information sources and marketing and/or economic information sources (e.g. *CIN, Chemical and Engineering News*). Technological decisions cannot be made without considering the corresponding marketing information (e.g. the replacement of a nickel catalyst in hydrogenation process by a cobalt catalyst depends, among other factors, on the present and projected price of the cobalt in comparison with that of nickel). One has to keep in mind that the projected and/or estimated marketing data are 'soft' information and, in contrast to scientific and/or technical data, cannot be verified experimentally. However, there are tools as well as expert market specialists and/or analysts that can help one with interpretations and projections. A chemist dealing with chemical marketing should use the various available tools and obtain a specialist's advice.

13.1. Chemicals prices

Some of the most important questions a manufacturer may be asking are who is the producer of the needed raw materials, how much is he charging, and what are his delivery terms? In contrast to the synthetic research chemist, who is interested in obtaining a chemical compound without paying too much attention to its price (a price of one dollar per gram or two dollars per gram does not make any difference to him), the manufacturer cares very much about the price he is being charged. A price difference of a few cents pre kilogram could make the difference between a profitable production line or a line that should be closed down. This kind of information concerning chemicals in bulk quantities can be found in the various buyer's guides.

The best buyer's guides available are the *SRI* (Stanford Research Institute) *International Directory of Chemical Producers — USA* and *SRI International Directory of Chemical Producers — Western Europe*. Two other good buyer's guides cover the USA — *The Chemical Week Buyers' Guide Issue* and the *OPD Chemical Buyers' Directory*. Although they do not contain all the information reported in the *SRI International Directory — USA*, they are much more readily available. There are also buyers' guides to other countries, e.g. Japan (*Japan Chemical Directory*), England (*Chemistry and Industry Buyers' Guide*). Two other guides that should be mentioned are the *European Chemical Buyers' Guide*, having an alphabetical listing of over 7000 chemicals and information on about 4000 suppliers, and *The Worldwide Chemical Directory*, which is an information list on the chemical industry throughout the world. Buyers' guides for research chemicals (*Fine Chemicals Directory*, etc.) have been discussed earlier (see p. 166).

Chemicals prices change very often and one should try and predict and/or forecast these changes. Furthermore, prices depend upon quantity and quality of the various compounds. Information about the current prices can be found in the various trade journals, e.g. *Chemical Marketing Reporter* gives week by week the current prices of chemicals and related materials. These are spot quotations and/or list prices of suppliers in New York or on some other indicated basis. It also reports US imports of chemicals and related materials. The same journal gives a weekly chemical profile of a particular basic chemical, giving information concerning its demand, supply, growth, price, usage, strengths and weaknesses, as well as a general outlook. Such a profile is extremely important when one is constructing a marketing report and/or forecast. One has to remember that all the information obtained from the *Chemical Marketing Reporter* covers primarily the USA. Similar information covering other countries can be found in other trade journals, e.g. *Japan Chemical Week* covers Japan, *European Chemical News* covers Western Europe.

One may construct and build a chemical compound profile using *CIN*, *Chemical Business NewsBase* and/or *Predicosts Overview of Marketing and Technology* (*PROMT*).

In 1980 the European Petrochemical Association (EPCA) created the EPCA Trade Statistics Data Base. The data base (which is available online via EDS International) contains international trade statistics for over 2000 chemicals. The data are organized by country and product, covering trading countries, products, value, volume, and cumulative volume and values.

Information about manufacturing plants, trade and production figures could be obtained online by using *Chem-Intell* or *Chemical Business News-*

base (*CBNS*). Both data bases are available on DATASTAR, Dialog and Pergamon Infoline. *Chem. Intell* is produced by Chemical Intelligence Service covering manufacturing plants as well as trade and production figures for over 100 organic and inorganic chemicals. *CBNS* is produced by the RSC giving worldwide chemical business news (with particular emphasis on European news) about chemicals usage and production.

13.2. Preparing a market study and/or forecast

A marketing study and/or forecast should take into consideration, in addition to the present situation, also the various factors influencing future sales and earning (e.g. price changes, antitrust laws and action, licensing agreements, joint ventures), future product demand, usage, and production (e.g. new capacity), as well as the prediction of government actions, business development, labour relations and financial aspects. One can locate such information with the aid of secondary sources such as *CIN* (see p. 57); *Predicast F&S Index Corporate Change*; *Predicast F&S Index International*; *Predicast F&S Index Europe* (The Predicast publications are compiled from over 2500 newspapers, trade journals, government documents, prospectuses, and reports from all around the world); etc.

There are some guides for the preparation of a market study and/or survey. One is K. Gorton and I. Carr's *Low Cost Market Research, A Guide for Small Businesses* (Wiley, 1983); another is *A Practical Guide to Market Research* published by the Bank Administration Institute (1983).

In many cases marketing reports are prepared externally by a consulting organization. The latest edition of *The Directory of US and Canadian Marketing Surveys and Services* edited by J. Ruch and published by Ruch Associates lists and describes the services of about 140 consulting firms (among them Arthur D. Little, Disclosure, Dun & Bradstreet, Predicast, Standard & Poor, SRI). This directory also serves as a reference guide to over 1500 multi-client reports which are commercially available. Another information source for commercially available multi-client reports and surveys is the latest edition of *Findex: The Directory of Market Research Reports, Studies and Surveys*, edited by S. De Gange and J. Verniero and published by FIND-SVP. Market research and industry survey reports held by the British Library could be located using the latest edition of *Market Research and Industry Surveys: A List of Reports Held by the Science Reference Library*.

In many cases the report is a custom designed one; such is the case especially when a multi-million dollar decision is dependent on it, which is the time when the best judgement and evaluation are needed. Such a custom made report not only covers in great depth the various marketing and technological aspects of the subject, but also includes field interviews with interested parties (producers, users, government officials, etc.) and in many cases it may even include computer modelling. Such a report is very expensive and may easily cost tens of thousands of dollars; however, this is a relatively small expense when one is dealing with a multi-million dollar decision.

13.3. Chemical processing and engineering information

Chemical processing information is difficult to obtain from the literature; very little is reported in the various chemical engineering journals, and whatever is reported, is reported in general terms, lacking any detailed description. Such detailed descriptions are the factors that determine the efficiency of a production line, and as efficient production is the key to profit making, these details are often kept as trade secrets.

One may obtain recent process technology data as well as evaluation and comparisons of various processes from specialized service sources. Two such sources out of many are: SRI International and Chem. Systems Inc. The services and reports of such organizations are expensive but the results justify such expense. Their product information includes reports on various chemical products production technology, evaluation of new technology developments and their significance, and information on the state of the art of process development.

An expert chemical engineer who knows the fine details of the particular technology may be able to obtain some of the data he needs from various encyclopaedias and reference books. Some of the useful single-volume treatises are the latest editions of J.A. Kent, *Riegel's Handbook of Industrial Chemistry* (Van Nostrandt, Reinhold, New York); F.A. Lowenheim and M.K. Moran, *Faith, Keyes and Clark's Industrial Chemicals* (Wiley, New York); R.N. Shreve and J.A. Brink, Jr., *Chemical Processes Industries* (McGraw-Hill, New York). Other important sources that should be used are the various chemical engineering encyclopaedias. The three best known ones are: *Kirk-Othmer Encyclopedia of Chemical Technology*; *Encyclopedia of Chemical Processing and Design* edited by J. McKetta; *Ullmann's Encyclopaedia of Industrial Chemistry*.

The third edition of the *Kirk-Othmer Encyclopedia of Chemical Technology* was published by Wiley in 24 volumes during the years 1978-84. The work covers chemical substances or classes of substances as well as fundumental principles and concepts, unit operation, analytical methods, and other topics. In 1985, after the completion of the third edition, the *Kirk-Othmer Concise Encyclopedia of Chemical Technology* was published. In this *Concise Encyclopedia* every one of the 1100 articles which appears in the *Kirk-Othmer* has been condensed and reviewed, either by the author or by another authoritative expert. Wiley also published the *Encyclopedia* Reprint Series which are reprint volumes containing articles grouped into specialized areas from the *Kirk-Othmer*. Some of the titles are: *Encyclopedia of Textiles, Fibers and Nonwovenfabrics*; *Encyclopedia of Semiconductor Technology*; *Lithium*; *Ground Water Quality*; *High Speed Fiber Spinning*. An online version of the *Kirk-Othmer* is available on BRS and Dialog. Recently a CD-ROM edition of the *Encyclopedia* became available to the public through Wiley Electronic Publication.

The *Encyclopedia of Chemical Processing and Design* is a completely new encyclopaedia edited by J. McKetta and published by M. Dekker. Its publication started at 1976 and up to the end of 1986 volumes have been published at the rate of 2–3 volumes a year. At this rate it will take many more years before the completion of the encyclopaedia. It is a comprehensive source of information on the design of equipments, systems, and controls utilized in chemical processing. The various topics are covered and discussed in depth—e.g. 132 pages are devoted to the topic of Catalytic Cracking.

The best known encyclopaedia is *Ullmann's Encyclopaedia of Industrial Chemistry*. The first four editions were published in German and the work is known as *Ullmann's Encyklopaedie der Technischem Chemie*. The fourth edition was published during the years 1972–84; immediately after its completion the publication of the fifth edition in the English Language began. Plans are for the publication of the *Encyclopaedia* in 36 volumes: 28 volumes constitute and alphabetically ordered compilation of articles about products and product groups in the chemical and allied industries, and 8 volumes provide information concerning basic knowledge on topics related to industrial chemistry (e.g. principles of chemical process engineering, analytical methods, environmental protection technology).

One may also use the general and/or specialized chemical engineering data bases for retrospective searches, e.g. *Chemical Engineering Abstracts, Theoretical, Chemical Engineering Abstracts, International Process Technology Abstracts*.

13.4. Bioengineering, biotechnology, and genetic engineering

Two of the most rapidly developing areas in science and technology during the last decade are the areas of bioengineering, and biotechnology and genetic engineering. Although these areas are biologically oriented, they are of much interest to the chemists and many chemists are involved in the various research and development projects in those areas.

There is a good information coverage of the period 1970–81 in the bibliography *Genetic Engineering, DNA, and Cloning, a Bibliography in the Future of Genetics* compiled by J. Menditto and D. Kirsch (Witston Publishing Co., Troy, N.Y., 1983). It includes books, scientific and newsy articles, and government reports. Another newsy type of information source in the area of genetic engineering is *Genetic Engineering News*, which is published by Mary Ann Liebert Inc., New York, ten times a year.

Few macroprofiles deal with bioengineering and its various topics. Engineering Information Inc. is publishing *Bioengineering Abstracts* monthly; BIOSIS is publishing *Bioresearch Today, Bioengineering and Instrumentation* monthly; and CAS is publishing four biweekly CAS BioTech Updates:

Biosensors, Environemental Biotechnology, Genetic Engineering, and *Pharmaceutical Application.*

A large number of specialized data bases provides bibliographic data and abstracts: Section 27 of Excerpta Medica— *Biophysics, Bio-Engineering and Medical Instrumentation*; Part 216 of *PASCAL Biotechnology* (it supercedes *Bulletin Signaletique* Part 215: *Biotechnology*). One should be aware of the various editions of *Bioengineering Current Awareness Notification (BECAN)* and use the appropriate one (e.g. *Biomechanics and Orthopaedics, Electrodes for Medicine and Biology*). Three data bases are devoted only to biotechnology and all of them are available online (although via different vendors): *Current Biotechnology Abstracts* (produced and published by the Royal Society of Chemistry), *Biotechnology Research Abstracts* (formerly *Biotechnology Abstracts*) (published and produced by Cambridge Scientific Abstracts); *Biotechnology Abstracts* (produced and published by Derwent).

Information about research projects, products, companies, and organizations in the biotechnology area could be found in the latest edition of the *International Biotechnology Directory* edited and compiled by J. Coombs. The directory covers topics such as biomass, diagnostics, fermentation, genetic engineering, industrial microbiology, oil recovery, pharmaceuticals, waste treatment, etc. Information sources in biotechnology could be easily located using A. Crafts-Lighty's guide *Information Sources in Biotechnology* which was published by Macmillan.

CHAPTER 14

Gathering Information about Individuals and Organizations

There are occasions when one is in need of some personal information concerning certain scientist or scientists (e.g. biography, affiliation, research interests). Sometimes the need is for information concerning a profit or non-profit organization (a research centre, an industrial complex, etc.). This required information may be a scientific one (e.g. research interests), personal one (e.g. list of scientists employed), or a commercial one (e.g. financial situation, marketing plans). Where and how can one obtain this kind of information?

14.1. Personal data

The first source one would think of using in order to locate somebody's biography is the national *Who's Who* (e.g. *Who's Who in America, Who's Who in Canada, Who's Who in Germany, Who's Who in Japan*). However, only a relatively small number of scientists' biographies are published in the various *Who's Who* books, owing to their selection policy (e.g. some of the guidelines for selecting scientists' biographies for *Who's Who in America* are '...selected memebers of the National Academy of Sciences, recipients of major national and international awards...').

The best place to locate an American or European chemist's biography is the latest edition of *American Men and Women of Science* (published by Bowker and available online via BRS and Dialog) or the latest edition of *Who is Who in Science in Europe* (published by Longman), respectively. One may use the Gale Research Company's latest edition and supplements *Biography and Genealogy Master Index* (it is available online via Dialog as *Biography Master Index*) in order to locate a biography source. It is a 'cumulated index' to over 600 various *Who's Whos*, bibliography dictionaries, etc.

There are some collections of biographies of notable chemists, e.g. D.M. Wynham, *American Chemists and Chemical Engineers* (published by the

American Chemical Society, 1976), E. Farber, *Great Chemists* (Interscience, 1961), and W.A. Campbell and N.N. Greenwood, *Contemporary British Chemists* (Taylor and Francis, 1971). However, one has to remember that these collections cover only a very small number of selected biographies.

The *CA* Author Index could be used in order to locate the affiliation or to determine the research interests of a particular chemist. The Author Index links principal authors (and patent assignees) to their publications via the *CA* Abstract Number (co-authors are cross-referenced to the principal authors). *CA* makes every effort to standardize authors' names by including their forenames in the paper version of the Author Index. However, in the online Author Index the author's name appears in exactly the way it was written in the original article; thus the same author may appear with his initials, with his forename(s) and/or initial and forename (e.g. Bergmann, E.D., Bergmann, Ernst D., and Bergmann, Ernst David, are three different name writings of the same author). Tha paper and the online versions of the Author Index are also different in their alphabetical arrangements. In the paper version the names are arranged alphabetically on the basis of their initials; thus Baker, Arthur David, appears before Baker, Alan Frederick, which appears before Baker A.L., which appears beford Baker, A.T. (A.D. before A.F. before A.L. before A.T.). In the online version the arrangement is purely alphabetical; thus the order of appearance of the four authors is Baker, A.L., Baker, A.T., Baker, Alan Frederick, Baker, Arthur David.

Turning to the abstracts themselves, one can reconstruct a list of publications, find out the author's affiliation, and after reading the abstracts, construct the research interest profile of the scientist in whom one is interested.

A source for obtaining lists of publications and research interests of chemists in academic institutions in the USA and Canada is the latest edition of the *Directory of Graduate Research*, published by the American Chemical Society (it is also available online via BRS). This is a guide to recent publications and current research interests of the faculty members in the graduate departments (those leading to a Ph.D. degree) of chemistry, chemical engineering, and biochemistry in the USA and Canada. The *Directory* is arranged alphabetically according to the various universities names. Thus, in order to be able to locate the desired information one has to know the academic affiliation of the person in whom one is interested. In cases where the affiliation is unknown one should first determine it by using the *CA* Author Index or the latest edition of the *National Faculty Directory* published by the Gale Research Company. The *National Faculty Directory* lists in alphabetical order all the faculty members of all junior colleges, colleges, and universities in the USA and their affiliations.

Recently (in 1984) the American Chemical Society published the *Chemical Research Faculty: An International Directory*. It gives information about faculty members in 735 chemistry, chemical engineering, biochemistry, and medicinal chemistry in over 60 countries. Although its coverage is not complete (country, institution, and department wise) it is a very valuable source of information about 8900 fculty members in some of the graduate departments around the world.

Good sources of current publications and research interests as well as personal data of individual chemists are the directories of their own institutions and/or departments, if available.

14.2. Profit and non-profit organizations

Information about the various activities and research goals of research centres, institutions, and organizations can be obtained by using their own directories. A general source for this type of information is the latest edition of *Research Centers Directory* and its periodical supplements, *New Research Centers*, published by the Gale Research Company. It is a directory to all university-related and other non-profit research organizations established on a permanent basis in the USA and Canada and carrying on continuing research programmes. Another Gale Research Company directory— *International Research Centers*. This directory covers government, university, independent non-profit and commercial research and development centers, institutes, laboratories, bureaus, test facilities, experimental stations, and data collection and/or analysis centres.

Information about research and/or development facilities and their parent companies in the USA could be found in the *Directory of American Research and Technology* (formerly entitled *Industrial Research Laboratories of the US*) published by Bowker and available online via Pergamon InfoLine. It covers about 11,000 facilities and laboratories giving personal information about their scientists and management as well as a description of their various research and development activities.

Longman has been publishing a large number of directories of various research centres which could help in looking for such information outside the United States—e.g. *European Research Centers*, *Materials Research Centers*, *Longman Guide to World Science and Technology* (a series of guides each devoted to a country or geographic region— *Science and Technology in the Middle East*, *Science and Technology in Latin America*, *Science and Technology in the United Kingdom*, *Science and Technology in China*, *Science and Technology in Japan*, etc.).

The best way to construct a general overview of an organization research and development activity is from the scientific publications and patents originating within that organization. The only paper product that links bibliographic citations of scientific articles to organizations in which the work was carried out is *Current Contents Address Directory*. The *Current Contents Address Directory* is an annual publication published in two parts: *Science and Technology* and *Arts and Humanities*. It is an ISI publication superceding the *Current Bibliographic Directory* which in turn superceded *Who is Publishing in Science*.

The *Current Contents Address Directory, Science and Technology*, is divided into three sections (or parts): an Author Section, an Organization Section, and a Geographic Section. The first section—the Author Section—lists the names and adresses of all publishing scientists and scholars whose work had been reported by any of the five *Current Contents* science and technology editions, followed by bibliographic citations of their work which was reported during the year in the various *Current Contents* editions. The address is given in the way it was reported in the original literature. This section can be used to obtain information concerning the affiliation of a publishing scientist and/or a list of his publications during the year. The Organization Section lists alphabetically the names and adresses of all academic, governmental, industrial, and any other organization throughout the world who had publishing scientists whose work had been reported in *Current Contents* during the year, followed by an alphabetical list of those publishing scientists. The Geographic Section lists alphabetically by state/country and organization the names of the organizations having publishing scientists. In order to reconstruct an organization's list of publications one has to look at the Organization Section to locate the publishing scientists of the specific organization he is interested in, and then turn to the Author Section to retrieve the publications of each of the publishing scientists.

When using the *Current Contents Directory Address* one has to bear in mind that the latest edition is not a replacement for the former one. It is not a cumulative publication index and each volume stands on its own as it describes what has happened during a particular year.

Current Contents, like any of the other ISI products, does not cover the patent literature, so information concerning industrial research and development is very limited (it is based only on the journal articles which were published by the industrial scientists, which is a minor communication channel of industrial research). The best information source of research profiles as well as trends in research and development of the various industrial companies and organizations is the secondary patent literature. One should use the INPADOC *Patent Application Service* (see p. 243), the *WPI* Patentee Index (see p. 244) or the *CA* Author Index (see p. 240) for that purpose.

A construction of research profile by online searching is much easier and simpler. Most data bases have a searchable corporate source field (e.g. *CA* on Dialog, Orbit, STN, etc., *Scisearch* on Dialog) or patent assignee (e.g. *WPI* on Dialog, Orbit, Questel, INPADOC on Pergamin InfoLine). Thus an online search in the appropriate field would give the desired information.

In many instances one is interested in a profit organization a corporation, an industrial plant, etc.) for reasons that have nothing to do with its scientific activity. Such is the case when one is interested in the physical location of the organization (principal and subsidiary offices and plants), its officers and executives (names, background, etc.), its acquisitions and merger history, its various production facilities, information about its products (present and future), its financial history (past, present, future), etc.

A lot of information can be found in the reports which public companies issue to their stockholders and/or file with Security and Exchange Commission(s) of the country(ies) in whose Stock Market their stocks are traded. Another source is the reports and annual compilations produced and published by various financial advisory companies (e.g. Arthur D. Little, Business Research Corporations, Disclosure, Dun & Bradstreet, Inter Company Comparisons Ltd. (ICC), Jordan & Sons Ltd., Media General Financial Services, Moddy's Investors Services, Predicast, Standard & Poor's Corporation, Trinet). The reports and the information are available in paper form as well as online. A source for localizing various annual reports and studies is *FINDEX: The Directory of Market Research Reports, Studies and Surveys* which is also available online via Dialog (File 196 *FIND/SVP*).

Current information covering topics such as production trends, construction and expansion of plants, novel processes and new products, corporate activities and news, and patent and licensing information are covered by *Biobusiness, Chemistry Business Newsbase, CIN*, and *PROMT* (see p. 250). Although these sources are primarily current awareness sources, they can and should be used for a retrospective searching in order to build up a general picture of the activity of a corporation.

CHAPTER 15

Expert Systems

About twenty years ago, J.M. Ziman states in his Book (J.M. Ziman, *Public Knowledge*, Cambridge University Press, Cambridge, 1968):

> The information system of science works quite well for the accumulation of details, but is failing in the equally essential task of assembling these details into a comprehensible analytically ordered and coherent system of ideas.

This problem—the conversion of information into knowledge—remained unchanged during the last two decades. J. Naisbitt has focused it very sharply in his book (J. Naisbitt, *Megatrends— Ten New Directions Transforming Our Lives*, Warner Books, New York, 1982):

> We are drowning in information but starved for knowledge.

In this book we have discussed the classical manual means of chemical information retrieval as well as the modern sophisticated computer systems that enable the chemist an interactive searching of large bibliographical, numerical, and chemical structures files. The ease with which such data can be stored and retrieved in machine-readable files has led to the development of computer systems which manipulate and correlate chemical data (textural, numerical, or structural) for other purposes than just information retrieval. One may look upon these systems as the first step of an automatic conversion of chemical information into knowledge. These computer systems are based upon the usage of one or more of the following techniques and/or methodologies: various statistical techniques, computer graphics, pattern recognition, artificial intelligence (AI), methodologies and programs, etc.

The most promising of all methods and techniques are the expert systems. There are various definitions for expert systems, one of the best ones being P.J. Deming's: 'An expert system is a computer system designed to simulate the problem solving idea of a human who is an expert in a narrow domain' [*Am. Sci.*, **74**, 18 (1986)]. A somewhat simple definition is 'computer system that incorporates human expertise'.

Expert systems are capable of solving problems that fall into one of the following categories—interpretation, prediction, diagnosis, debugging, design, planning, monitoring, repair, instruction, and control. They are performing tasks at an expert level of performance, emphasizing domain-specific problem-solving strategies, and employing self-knowledge to reason their own inference processes and provide an explanation and/or justification for their reached conclusions.

An expert system is composed of three parts: a data base (knowledge base); algorithm(s) for constructing proofs (inference engine); and a user interface which provides the connection between the user and the system.

One can buy today on the open market an expert system, 'shells'. These systems come with empty data bases; thus one has to build the desired data base for a specific system. The data base building is carried out by a trial and error process known as knowledge engineering (gathering and executing the knowledge and experience of human experts). Expert systems can be divided into shallow and deep systems. The distinction between those two types of systems is not sharp and some think that it is even not important. Shallow systems have limited data bases which store mainly facts and a very limited number of rules. Their conclusions result from very short proofs and usually are a straightforward result of the information included in the data bases. In many cases they give unsatisfactory results to problems other than those for which they have been tested. Deep systems contain a large (sometimes very large) number of rules as well as various models and their first principles. Their conclusions are derived from long proofs and in many cases may be anything but obvious.

Various chemical expert systems are in existence and operation today dealing with synthesis design, structure determination, structure-activity relationship, etc. Others are in various stages of design formation, testing (e.g. separation science, safety).

All expert systems can be divided into two general types: information-oriented systems and logic-oriented systems. Information-oriented systems are empirical by nature; they depend on the usage of algorithms that process the input data into specific prescribed ends. On the other hand the logic-oriented systems are essentially non-empirical; they are based on rigorous mathematical structure and formalism. The main problem with the application of the logic-oriented systems appears to be the translation of the problem and the data sets (which are usually too big and too heterogeneous) into a pure mathematical form.

15.1. Computer-aided synthesis design

Developing a system that will assist the synthetic chemist in designing an organic synthesis is quite a challenge. Such a system should be able to search for an effective strategy for the simplification of the synthetic problem as well as being able at the same time to generate a large number of pathways of synthetic steps connecting the target molecule with the starting materials.

There are two different directions of a synthetic analysis; the direction referred to as 'synthetic', which corresponds to the direction of the execution of the synthesis, and the direction referred to as 'retrosynthetic', in which the target structure is transformed in a number of steps through simpler structures into a starting material(s) for the synthesis. There are different terminologies for synthetic and retrosynthetic analysis as one can see below.

Direction of laboratory execution is 'synthesis', represented by a single arrow (→) and the process is called a 'reaction'. The reaction is keyed by reactive groups in the starting material that activates the molecule to undergo a synthetic step.

Direction of the logical analysis is 'retrosynthesis', represented by a double arrow (⇒); and the process is called a 'transform'. The transform is keyed by critical subunits (known as retron) that 'activate' the application of the transform to the target structure.

Within the last two decades a number of synthesis design programs has been developed in various academic and research institutions. Some of the well known programs are: *LHASA* (*SECS* and *CASP* are derivatives of *LHASA*), *SYNCHEM*, *CAMEO*, *SYNGEN*, and *EROS*. Some of the programs are empirical ones which make use of the reactions library, others are non-empirical making use of structural transformation. While some of the programs are retrosynthetic, others are synthetic ones. There are programs which are interactive, thus enabling the synthetic chemist to guide the selection of promising routes; others are executed without any human intervention.

15.1.1. *LHASA* (Logic and Heuristics Applied to Synthetic Analysis)

LHASA has been developed by Corey and his group at Harvard during the last two decades. It is an interactive system using graphical images (which is the natural language of the chemist) for chemist–machine interaction. The system is based on a data base of well over a thousand reactions, with the chemist contributing his experience and intuition. *LHASA* is taking the retrosynthetic analysis approach generation retrons. The analysis proceeds via one or more transform strategies: 'simplifying transforms' (they correspond to powerful synthetic reactions—reactions in which rings or ring pairs are formed—e.g. Diels–Alder, Rubinson annulation, various cyclizations, sigmatropic rearrangements); 'mechanistic transformation' (they are the retrosynthetic version of step by step electronic mechanisms); 'topological strategies' (they identify one or more bonds whose breakage would lead to a big molecular simplification); 'stereochemical strategies' (they are heuristically derived methods for reducing stereochemical complexity in the retrosynthetic direction); and "functional group strategies" (they are derived from synthetic experience and used for simplifying the various transforms).

A detailed description of the LHASA could be found in:

(1) E.J. Corey, A.K. Long, and S.D. Rubenstein, *Science*, **228**, 408 (1985).
(2) A.K. Long, S.D. Rubenstein, and L.J. Joncas, *Chem. Eng. News*, **61**(19), 22 (1983).

A similar system which one may look upon as a derivative of *LHASA* is *SECS* (Simulation and Evaluation of Chemical Synthesis). It was developed by Wipke and his coworkers (Wipke was involved in the early stages of the LHASA development) at the University of California, Santa Cruz. Its main advantage and/or difference over the original *LHASA* program is that it includes strong stereochemical consideration. Its aim is synthetic stereochemistry and capable of three-dimensional modelling using energy minimization routines. The original version of *LHASA* did not have any stereochemical considerations; however, later versions contain stereochemical strategies.

An offspring of *SECS* which has been developed in Europe by a consortium of Swiss and West German pharmaceutical companies is *CASP*.

More information concerning SECS could be found in:

(1) W.T. Wipke, G.I. Ouchi, and S. Krishman, *Artif. Intelligence*, **9**, 173 (1978).
(2) W.T. Wipke, H. Brawn, G. Smith, F. Choplin, and W. Sieber, *ACS Symp. Ser.*, **61** (*Comput. Assisted Org. Synth.*), 97 (1977).

15.1.2. *SYNCHEM* (SYNthetic CHEMistry)

This program has been developed by H. Gelernter and his group at the State University of New York, Stony Brook. It is similar to the *LHASA* in that it uses large data bases of chemical reactions and starting materials. However, it is a non-interactive program. The heuristics needed to decide which of the precursors are the ones for further elaboration are included within the system, and the system output contains all proposed synthetic routes to the desired molecule. It has been introduced to the system by a training process based upon S. Warren's book '*Designing Organic Syntheses, A Programmed Introduction to the Synthetic Approach*' (Wiley, New York, 1978). *SYNCHEM*, contrary to *LHASA*, is capable of operating not only in the retrosynthetic route but in the synthetic route as well; thus it is able to generate products from a given set of reactants through a specified sequence of reactions.

Further information concerning the SYNCHEM program could be found in:

(1) H. Gelernter, S.S. Bhagwat, D.L. Larsen, and G.A. Miller, *Anal. Chem. Sym. Ser.*, **15** (*Comput. Appl. Chem.*), 35 (1983).
(2) K.K. Agarwal, D.L. Larsen, and H.L. Gelernter, *Comput. Chem.*, **2**, 75 (1978).

15.1.3. *CAMEO*

CAMEO is one of the non-empirical approaches to organic synthetic design. It predicts the products of organic reactions given the starting materials and conditions. The prediction is not data driven, but is made on the basis of mechanistic considerations and operates in the synthetic direction. It mimics the logical processes of a synthetic organic chemistry as closely as possible.

The system is being developed by W.L. Jorgensen at Purdue University. At present the system contains modules for: base catalysed reactions, nucleophilic reactions, acid catalysed reactions, nucleophilic aromatic substitutions, electrophilic aromatic substitutions, selected organometallic reactions, reactions of organosilicon compounds, pericyclic reactions (cycloadditions, sigmatropic and electrocyclic rearrangements). *CAMEO* is capable of carrying out a stereochemical analysis, thus being able to ascertain stereochemical and/or physical restriction on the synthetic reaction.

Detailed information on the system could be found in the following references:

(1) M.G. Bures, B.L. Roos–Kozel, and W.L. Jorgensen, *J. Org. Chem.*, **50**, 4490 (1985).
(2) C.E. Peishoff and W.L. Jorgensen, *J. Org. Chem.*, **50**, 3174 (1985).
(3) C.E. Peishoff and W.L. Jorgensen, *J. Org. Chem.*, **50**, 1056 (1985).
(4) J.S. Brunier and W.L. Jorgensen, *J. Org. Chem.*, **49**, 3001 (1984).
(5) J.A. Schmidt and W.L. Jorgensen, *J. Org. Chem.*, **48**, 3923 (1983).
(6) C.E. Peishoff and W.L. Jorgensen, *J. Org. Chem.*, **48**, 1970 (1983)
(7) T.D. Salatin, D. McLaughlin, and W.L. Jorgensen, *J. Org. Chem.*, **46**, 5284 (1981).
(8) T.D. Salatin and W.L. Jorgensen, *J. Org. Chem.*, **45**, 2043 (1980).

The difference between *CAMEO* and *LHASA* or *SYNCHEM* could be summarized as follows: *CAMEO* tells the chemist, using detailed mechanistic considerations, if subjection of compound X to reaction conditions Y will result in the formation of compound Z or maybe of something else, while *LHASA* and *SYNCHEM* help the chemist to decide which reagents and precursors should be used for the synthesis of compound Z.

15.1.4. *EROS* (Elaboration of Reactions for Organic Synthesis)

EROS is another non-empirical synthesis program. It has the capability of proceeding in either the retrosynthetic or the synthetic route. It is based on using a mathematical model of constitutional chemistry. In this model each molecule is represented in the form of a BE (bond/electron) matrix. Reactions are represented by matrix differences which describe the redistribution of valence electrons which took place during the reaction. For each precursor the program calculates the reaction energy which is needed for that transformation, ranking the precursors on this basis. One may look upon *CHIRP* as an offspring of *EROS*. It aims at industrial chemical reactions using the Gibbs free energy as a feasibility measurement.

Detailed information on *EROS* and *CHIRP* could be found in the following references:

(1) J. Gasteiger, *Comput. Aided Mol. Des.*, p. 197 [Oyez. Sci. Tech. Serv. London (1985)] *CA*, **104**, 129110f (1986).
(2) R.B. Agnihotri and R.L. Motard, *ACS Sym. Ser.*, **124**, (*Comput. Appl. Chem. Eng.*) 193 (1980).
(3) J. Gasteiger, C. Jochum, M. Marsili, and J. Thoma, *MATCH*, **6**, 177 (1979).
(4) J. Gasteiger and C. Jochum, *Top. Curr. Chem.*, **74**, 93 (1978).

15.1.5. *SYNGEN*

SYNGEN was developed at Brandeis University by Hendrickson and his group. It differs from all other systems and programs in that its goal is to obtain efficient synthetic routes by converging as rapidly as possible on the starting material. This is done by reducing to an absolute minimum the number of intermediates generated during the synthesis. It is done by seeking whole routes between starting materials and targets rather than building reaction trees.

Full information on the system protocol and structure is to be found in:

(1) J.B. Hendrickson, *Acc. Chem. Res.*, **19**, 274 (1986).
(2) J.B. Hendrickson, *J. Am. Chem. Soc.*, **108**, 6748 (1986).
(3) J.B. Hendrickson, D.L. Grier, and A.G. Toczko, *J. Am. Chem. Soc.*, **107**, 5228 (1985).
(4) J.B. Hendrickson, E. Braun-Keller, and G.A. Toczko, *Tetrahedron Suppl.*, **9**, 359 (1981).

15.2. Computer-aided structure determination

In the last 10–15 years the applications of computer methods for the identification of complex mixtures of known compounds (pollutants, body fluids, overdose victims, etc.) as well as structure determination of unknown chemical compounds (from simple organic compounds to complex natural products) has been increased dramatically. The usage of mass spectrometry for the identification of complex mixtures has been discussed recently [F.C. McLafferty and D.S. Stauffer, *J. Chem. Inf. Comput. Sci.*, **25**, 245 (1985)]. A general overview of the various problems and solutions to computer-aided structure determination is given in N.A.B. Gray's book *Computer-Assisted Structure Elucidation* (Wiley, Chichester, 1986, 550pp).

The oldest and best known computerized structure elucidation program is

266

DENDRAL (DENDRitic ALgorithm for acyclic isomer generation). *DEN-DRAL* has been developed over many years at Stanford University. The whole project has been summarized in a book: *Application of Artificial Intelligence for Organic Chemistry: The DENDRAL Project*, by R.K. Lindsay, B.G. Buchanan, E.A. Feigenbaum, and J. Lederberg (McGraw Hill, New York, 1980). Recent publications dealing with the application of the *DENDRAL* program are:

(1) M.R. Lindley, J.N. Shoolery, D.H. Smith, and C. Djerassi, *Org. Magn. Reson.*, **21**, 405 (1983).
(2) C. Djerassi, D.H. Smith, C.W. Crandell, N.A.B. Gray, G.J. Nourse, and M.R. Lindley, *Pure Appl. Chem.*, **54**, 2425 (1982).
(3) D.H. Smith, N.A.B. Gray, J.G. Nourse, and C.W. Crandell, *Anal. Chim. Acta*, **133**, 471 (1981).
(4) J. Finer-Moore, N.V. Mody, S.W. Pelletier, N.A.B. Gray, C.W. Crandell, and D.H. Smith, *J. Org.Chem.*, **46**, 3399 (1981).
(5) R.E. Carhart, D.H. Smith, N.A.B. Gray, J.G. Nourse, and C. Djerassi, *J. Org. Chem.*, **46**, 1708 (1981).

Various other programs for structure generation have been developed (e.g. Munk's *CASE*, Sasaki's *CHEMICS*, Gribov's *STREC*). More information on *CHEMICS* (Combined Handling of Elucidation Methods for Interpretive Chemical Structures) could be found in:

(1) S. Sasaki and Y. Kudo, *J. Chem. Inf. Comput. Sci.*, **25**, 252 (1985).
(2) S. Sasaki and H. Abe, *Anal. Chem. Symp. Ser.*, **15**, (*Comput. Appl. Chem.*), 185 (1983).

15.3.   Computer-assisted structure activity relationship (SAR)

SAR is important in areas such as pharmacology—drug design; agriculture—herbicides and/or pesticides design; food industry—flavour design; toxicology and/or genetics—carcinogenicity, mutagenicity. SAR is a correlation based on pharmacology, biochemistry, organic chemistry, physical chemistry, and quantum chemistry. It is a study on a molecular level of activity and/or mechanisms. Such a study could benefit from the usage of computer-assisted correlative methods for relationships from tested molecular structures. Indeed, various approaches to computerized structure-activity studies have been developed during the last few years. Most developments are taking part in the area of computer-assisted drug design. Recent advances and perspectives have been reported—A.J. Hopfinger, *J. Med. Chem.*, **28**, 1133 (1985).

CHAPTER 16

Perspectives

During the last decade computer based information systems replaced the traditional information systems (which were based on paper products). Computer based chemical information systems are better developed and offer much more sophisticated means of retrieval than information systems in any other area of science. This is caused by the availability of bibliographic, numeric, and structural data bases and their integration into one information system. However, there is much more to come—e.g. Markus DARC, a generic compound search against generic compound data base; storage and retrieval of text and graphic in one data base, the German Patent Database project carried out by the German Patent Office; and FIZ Karlsruhe; the automatic conversion of chemical names (not necessarily *CA* Index Names) into structural formulae.

With the rapidly changing technological picture it is impossible to predict accurately the future of the chemical information systems (and that of information systems in general) nor that of the science library. One thing is clear, chemical information systems will become more and more sophisticated with the application of the recent developments in computer hardware and software, storage, computer communication, and various related areas. New modes and techniques for handling, synthesizing, organizing, presenting, and dissimilating information will emerge in the near future.

The number of scientific journals will decrease; some journals will become electronic ones, others will be available in an electronic form as well as in paper and/or micro form (the scientific journals published by the ACS, RSC and Wiley are available today online via STN). Most personal subscriptions will be replaced by selective distribution of scientific articles. It is interesting to know that during the last decade the reading of journal articles from library materials has increased. The increase is mainly in current articles found by browsing in recently published journal issues. This suggests that libraries will continue buying current journals and store them in paper form. However, old material would be stored in an electronic form [probably on compact optical discs (CD ROM) or 12 inch optical discs]. Electronic delivery systems will replace interlibrary lending. New breakthroughs that would come from developments in artificial intelligence will permit computerized and/or computer-aided translations in real time.

One has to realize that information is today considered (and will be considered even more so in the future) a commodity and as such it is not free. However, the end user who pays for the information retrieved would like to get maximum recall and precision. R. Luckenbach, the director of the Beilstein Institute, defined the problem, showed the reasons, and suggested a solution in an editorial published in the November 1984 issue of *Accounts of Chemical Research*.

'...The problem of information retrieval in organic chemistry has changed with the growth in complexity of the science, and with the development of new retrieval tools. However, one fundamental aspect remains unchanged: the user of an information system hopes to obtain relevant answers to his questions, but with as little ballast as possible. This ideal requirement must be viewed against the background of two factors:

(I) The rapid growth of research and development in all areas of chemistry in the last 20 years has led to such a flood of published material that no single worker can cope with the sheer amount of data, let alone maintain a critical evaluation of all that may be relevant to his work.

(II) The amount of information published is strongly influenced by socioeconomic factors, which, however, appear to exert less influence on the quality of the work...'

'...The nonselective storage of all incoming information in computerized retrieval systems does not remove these problems. The percentage of ballast produced often forces the user to make an extremely time-intensive examination of the indicated literature, whereby the time required for the organization aspects of the appraisal (ordering/finding the original, translation, locating the supposedly relevant entry in a mass of discussion or experimental details) far outweighs the time required for his critical evaluation of the actual hard content sought for.

Since this critical screening on any particular subject has to be carried out afresh by every individual seeking information, the entire process must be viewed as making nonsense of the concept of economy of effort.

It would thus appear meaningful and rational not to leave it to each individual end user to rid the material of balast himself but instead to make the information available after critical screening and assessment by a competent 'information analysis center'.'

Indeed the various producers of chemical information began recently screening, analysing, and manipulating their data bases and this tendency will be growing in the coming future. ISI is planning to introduce its Chemistry Citation Index (CCI), identifying research fronts in chemistry and producing chemistry volumes in their *Atlas of Science* (see p. 68); two of the projected volumes are: *Atlas of Chromatography* and *Atlas of Organic Chemistry*. CAS is working on the production of two new data bases: Patent data base and Chemical Reaction data base.

We shall be seeing in the future increasing integration of services and datafiles. Some integrated systems are available today (CIS; Telesystems-DARC, CAS ONLINE Files— *CA, CAOLD, REG*) while others are in planning and preparation (e.g. the incorporation of the Patent data base and the Reaction data base into CAS ONLINE).

The information producers want more and more to get involved not only with data base production but with its distribution as well (distribution of the printed and/or electronic forms). Already today some of the data base products are available through the producers only (e.g. *CA* abstracts, *CAOLD* file, *CA Preview* file, *CASREACT* file and pre-1965 registration compounds are available online only through STN International). The direct producers involvement with the end user is going to increase in the future, which means that CAS is going to be the dominant force in marketing and distributing chemical information.

There is going to be a movement back to the chemist (or the end user) as the operator of the computerized chemical information systems. The major online information vendors and producers have started to promote the usage of their systems by the chemists themselves (e.g. CAS ONLINE is giving workshops for chemists—'*CA* File for Chemists'—and holding a 'Train the Trainer' program). Search systems for the inexperienced searcher are becoming available at present. Some of the vendors have recently started menu-driven search systems—Dialog's *Knowledge Index*, BRS's *After Dark*. Those systems contain many chemical data bases that can be searched by the chemist during off-peak hours at reduced prices (e.g. BRS's *After Dark* contains the following chemistry data files: *Agricola BIOSIS, CA Search, Kirk–Othmer/* Online, *MEDLINE, NTIS*). Software searching packages are becoming available; those packages help the searcher to conduct his search (e.g. *SCIMATE* produced by the ISI enable one to search *SciSearch* and *MEDLINE*). More and more menu-driven search systems for the inexperienced and/or the infrequent searcher will become available. One such system which has been operating for some time is EASYNET. It is connected to sixteen vendors (among them BRS, DIALOG Pergamon Infoline, Questel, Orbit) covering over 600 different data bases, many of which are of interest to chemists.

** EASYNET **
The Knowledge Gateway
$8 per search, 20 cents per minute
telecommunications charge.
Please enter your Visa, Mastercard,
American Express or EasyNet Account
number.
Type L to leave EasyNet.

270

→ ∗∗∗∗∗∗∗∗∗∗∗∗∗
Card Holder's Name:
→ **Y.WOLMAN**

∗∗ **EasyNet Account** ∗∗
∗∗∗∗∗∗∗∗∗∗∗∗∗∗

Y.WOLMAN

Is this correct? (Yes/No) → Y
Please standby while we approve your credit.

```
+ - - - - - - - - - - - - - - - - - - - - - - - - - - - - - - - - - +
!                                                                   !
! WELCOME TO EASYNET                                                !
! (c) 1986 Pat. Pend.                                               !
! Telebase Systems Inc.                                             !
! Customer Service                                                  !
! 1-800-841-9553                                                    !
! (in PA 215-296-1793)                                              !
!                                                                   !
+ - - - - - - - - - - - - - - - - - - - - - - - - - - - - - - - - - +
```

Logon JUL 31 , 1986 at 5:50

PRESS   TO SELECT
  1   EasyNet-I   We pick the database
  2   EasyNet-II   You pick the database
  H   Help
Total Charges thus far   : $0.20
→ **1**

PRESS   TO SELECT
  1   Subject
  2   Person
  3   Place
  4   Organization
  H   Help
Total charges thus far   : $0.40
→ **1**

PRESS   TO SELECT
  1   **Current Events**
  2   **Business, Economics**
  3   **Computers, Sci/Tech, Medicine**
  4   **Law, Trademarks, Patents**

5	Social Sciences, Education
6	Art, Literature, Entertainment
7	Religion, Philosophy
H	Help

Total charges thus far : $0.40

→ 3

PRESS	TO SELECT
1	Agriculture
2	Biology
3	Chemistry
4	Computer, engineering, technology
5	Earth sciences, energy
6	Mathematics, physics
7	Medicine, Allied Health
H	help

Total charges thus far : $0.40

→ 3

PRESS	TO SELECT
1	Research journals and patents
2	Lab Hazards Bulletin
3	Trade journals
4	Popular magazines
5	Books on chemistry
6	Encyclopedias
7	List of related databases
H	Help

Total charges thus far : $0.40

→ 7

PRESS	TO SELECT
1	CHEMICAL ABSTRACTS
	Comprehensive worldwide coverage
2	CHEMICAL ENGINEERING ABSTRACTS
	Theoretical and practical aspects
3	CHEMICAL EXPOSURE
	Toxicology of chemicals
4	CHEMICAL INDUSTRY NOTES
	Business of chemical industries
5	CHEMICAL REGULATIONS & GUIDELINES
	US Government statutes, etc.
6	CONFERENCE PAPERS INDEX
	Papers of conferences worldwide
7	other databases

H   Help
Total charges thus far   : $0.40
→ 7

PRESS   TO SELECT
  1   DISSERTATION ABSTRACTS ONLINE
                         Doctoral (some Masters theses)
  2   FINE CHEMICALS DIRECTORY
                         Suppliers of chemical substances
  3   FOOD SCIENCE AND TECHNOLOGY
                         Food processing, packaging
  4   INDEX CHEMICUS ONLINE
                         Info. on new organic compounds
  5   JANSSEN
                         Organic chemical product catalog
  6   KIRK-OTHMER ONLINE
                         Encyclopedia of chemical tech.
  7   other databases
  H   Help
Total charges thus far   : $0.60
→ 7

PRESS   TO SELECT
  1 NTIS
          Government sponsored research
  2 SCISEARCH

                               Multi-disciplinary research
  3 7TSCA INITIAL INVENTORY

                           Commercial chemicals in US
  4 previous choices
  HHelp
Total charges thus far   : $0.60
→ 1

Enter your specific topic.
(type H for important examples)
  or B to back up)
→ H

CONNECTING WORDS
  Don't use small words like:
     by, from, in, of, the, at
  EX: Joan Arc instead of
     Joan of Arc.
WILD LETTERS

Use / as a 'wild letter' at the
    end of a word.
Ex: democ/ will retrieve democracy
                            democratic
                            Democrats
Tax/ will retrieve tax
                    taxes
                    taxation
Press (return) to continue... →

LOGIC WORDS (and, or, not)
    Use   AND   to find items common
        to two or more subjects.
    Ex: dog AND leash
        police AND civilian control
        debt AND management
    Use OR to find items on
        either or both subjects.
    Ex: Bach OR Handel
        dog OR cat OR pet
Press (return) to continue... →

    Use   NOT   to exclude a
        subject from another.
    EX: candy NOT taffy
        housing NOT mobile homes
    Use ( ) around groups.
    EX: (dog OR cat OR pet) AND leash
        eskimo/ AND (lawyer/ or attorn/)
Press (return) to continue... →

Enter your specific topic.
            *
            *
            *
→ L
    Charges:

System Access:
    Telecommunications $1.80
Database Charges:
    1 Searches: $8.00
    0 Reprints: $0.00
    0 Express Reprints: $0.00
    0 Abstracts $0.00

| Surcharges | $0.00 |
| Total Charges: | $9.80 |

Logoff 0208029 31JUL86 5:59 EST

Thank-you for calling !
Goodbye, Please hang up now.

An experimental intelligent gateway computer is in operation in the TIS (Technology Information System) at Lawrence Livermore National Laboratory. This gateway is capable of performing an automatic access to external data resources (e.g. the vendor's host computer) via computer communication networks and/or direct telephone dialup. It is also capable of modelling and analysing the data retrieved as well as transferring the original and/or workup data anywhere in the world. Some type of intelligent gateway will probably be incorporated in one way or another in the library of the future.

More and more expert systems will become available and will be used by the chemist. The role of the information specialist is going to change again, as most of the online searches are going to be carried out by the end user from his office or laboratory. The information specialist will become involved with the chemists' research problems, advise them on the existence and applications of the various information systems, create and control data bases for inhouse usage, and perform complicated and/or specialized online searches. His new duties will include teaching and directing the chemist on how to maximize intelligent use of the data retrieved as well as helping the user to perform the various steps in the process of data treatment.

An interesting article devoted tos the perspectives of information transfer has been published recently by C.M. Bowman, J.A. Nosal and A.E. Rogers— Effect of New Technology on Information Transfer in the 1990s. *J. Chem. Inf. Comput. Sci.*, **27**, 147 (1987).

Glossary of Acronyms

ABS	Abstract
ACS	American Chemical Society
AMP	Adenosine-5-phosphate
API	American Petroleum Institute
APIP-44	American Petroleum Institute Project 44
ASCA	Automatic Subject Citation Alert
ASCII	American Standard Code for Information Interchange
ASTM	American Society for Testing and Materials
BA	*Biological Abstracts*
BA/RRM	*Biological Abstracts/Reports, Review, Meetings*
BECAN	Bioengineering Current Awareness Notification
BIB	Bibliography
BIOSIS	Biosciences Information Services
BLDSC	British Library Document Supply Centre
BLLD	British Library Lending Division
BRS	Bibliographic Retrieval Services
BSC	British Safety Council
C13 NMR	^{13}C NMR Data Base
CA	*Chemical Abstracts*
CAB	Commonwealth Agricultural Bureaux
CAC	*Current Abstracts of Chemistry*
CAC&IC	*Current Abstracts of Chemistry and Index Chemicus*
CAN	Chemical Abstracts Number
CARI	*Chemical Abstracts Review Index*
CAS	Chemical Abstracts Service
CAS-RN	Chemical Abstracts Service Registry Number
CASDDS	Chemical Abstracts Service Document Delivery Service
CASSI	*Chemical Abstracts Service Source Index*
CBNS	*Chemical Business Newssbase*
CCR	*Current Chemical Reactions*
CCS	Coordination Compound
CD	Circular

CD ROM	Compact Disc, Read Only Memory
CEA	*Chemical Engineering Abstracts*
CHEMICS	Combined Handling of Elucidation Methods for Interpretive Chemical Structures
CI	Class Identification
CI	Collective Index
CIN	*Chemical Industry Notes*
CINDAS	Centre for Information and Numerical Data Analysis and Synthesis
CIS	Centre International d'Information et d'Hygiene du Travail
CIS	Chemical Information System
CISTI	Canadian Institute for Scientific and Technical Information
CN	Chemical Name
CNMR	^{13}C NMR Search System
CODATA	Committee on Data for Science and Technology
CODEN	Code Designate
COM	Component
COM	Computer on Microfiche
CPI	*Central Patent Index*
CPI	*Chemical Patent Index*
CRC	Chemical Rubber Company
CRDS	Chemical Reactions Documentation System
CRT	Cathode Ray Tube
CSI	*Chemical Substance Index*
CTRC	Colorado Technical Reference Center
DARC	Description, Acquisition, Retrieval, and Correlation
DDC	Dewy Decimal Classification
DDS	Document Delivery Service
DELFTLIB	Delft University of Technology Library
DENDRAL	Dendoritic Algorithm for acyclic isomer generation
DMF	Dimethylformamide
DOC	*Dictionary of Organic Compounds*
EI	Engineering Index
EIC	Environmental Information Centre
EMBASE	Excerpta Medica Base
EPA	Environmental Protection Agency
EPCA	European Petrochemical Association
EPI	Electrical Patent Index
EPO	European Patent Organization
EROS	Elaboration of Reactions for Organic Synthesis
ESA	European Space Agency

ESA-IRS	European Space Agency—Information Retrieval Service
FACT	Facility for Analysis of Chemical Thermodynamics
FCD	*Fine Chemical Directory*
FEBS	Federation of the European Biochemical Societies
FEDRIP	Federal Research in Progress
FID	International Federation of Documentation
FIZ	German Information Centre for Energy, Physics and Mathematics
FR	Formula of Rings
GC	Gas Chromatography
HEEP	Health Effects of Environmental Pollution
IBRS	*Index to Book Reviews in the Sciences*
IC	*Index Chemicus*
ICC	Intercompany Comparisons Ltd.
ICI	Information Consultants Inc.
ICSU	International Council of Scientific Unions
ICT	*International Critical Tables*
INFO	Information on Demand
INPADOC	International Patent Documentation Centre
INFODOC	Information/Documentation
IPC	International Patent Classification
IR	Infrared
IRAC	*Index of Reviews in Analytical Chemistry*
IRC	International Research Council
IRIS	Infrared Information System
IROC	*Index of Reviews in Organic Chemistry*
ISBN	International System Book Number
ISI	Institute of Scientific Information
ISR	*Index to Scientific Review*
ISSN	International System Serial Number
ISTP	*Index to Scientific and Technical Proceedings*
ITC	International Translation Center
IUPAC	International Union of Pure and Applied Chemistry
JICST	Japan Information Center of Science and Technology
KWIC	Keywords in Context
LC	Library of Congress
LC	Liquid Chromatography
LHASA	Logic and Heuristics Applied to Synthetic Analysis

MACCS	Molecular Access System
MCA	Manufacturing Chemists' Association
MF	Molecular Formula
NBS	National Bureau of Standards
N/H	National Institute of Health
NIOSH	National Institute for Occupational Safety and Health
NMR	Nuclear Magnetic Resonance
NRCC	National Research Council of Canada
NSRDS	National Standard Reference Data System
NTIS	National Technical Information Service
OATS	Original Article Tear Sheet
ORAC	Organic Reactions Accessed by Computer
OSRD	Office of Standard Reference Data
PC	Personal Computer
PCF	Parent Compound File
PCI	Parent Compound Index
PCT	Patent Cooperation Treaty
PMR	Proton Magnetic Resonance
PROMT	Predicasts Overview of Marketing and Technology
PSI	Permuterm Subject Index
REACCS	Reaction Access System
REG	Registry
RF	Ring Formula
RFI	Ring Formula Index
RN	Registry Number
ROSPA	Royal Society for the Prevention of Accidents
RSC	Royal Society of Chemistry
RSF	Ring System File
RSI	Ring Substructure Index
RTECS	Registry of Toxic Effects of Chemical Substances
SANDRA	Structure and Reference Analyser
SANSS	Structure and Nomenclature Search System
SCI	Science Citation Index
SDI	Selective Dissemination of Information
SECS	Simulation and Evaluation of Chemical Synthesis
SEMT	Science, Engineering, Medicine, and Technology
SIGLE	System for Information on Grey Literature in Europe
SRI	Stanford Research Institute
SRIM	Selected Research in Microfiche
SSIE	Smithsonian Science Information Exchange

STN	Scientific and Technical Information Network
SY	Synonym
SYN CHEM	Synthetic Chemistry
THF	Tetrahydrofuran
TIS	Technology Information System
TPRC	Thermophysical Research Data Center
TRC	Thermodynamic Research Center
UDC	Universal Decimal Classification
UMI	University Microfilm International
UMIACH	University Microfilm International Article Clearing House
UV	Ultraviolet
VINITI	All Union Institute for Scientific and Technical Information
VTTINF	Technical Research Center of Finland
WIPO	World Intellectual Property Organization
WLN	Wiswesser Line Notation
WPA	*World Patent Abstracts*
WPI	*World Patent Index*

Index

286